The NEC
Compared and
Contrasted

The NEC
Compared and
Contrasted

Edited by

Frances Forward

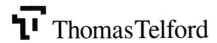 Thomas Telford

Published by Thomas Telford Publishing, Thomas Telford Ltd, 1 Heron Quay,
London E14 4JD. URL: http://www.thomastelford.com

Distributors for Thomas Telford books are
USA: ASCE Press, 1801 Alexander Bell Drive, Reston, VA 20191-4400, USA
Japan: Maruzen Co. Ltd, Book Department, 3–10 Nihonbashi 2-chome, Chuo-ku,
    Tokyo 103
Australia: DA Books and Journals, 648 Whitehorse Road, Mitcham 3132, Victoria

First published 2002
Reprinted June 2003
Reprinted March 2004

Also available from Thomas Telford Books
The New Engineering Contract, 2nd edition. Institution of Civil Engineers.
    ISBN 07277 2094 5
Civil Engineering Construction Contracts, 2nd edition. M O'Reilly. ISBN 07277 2785 0
The NEC Engineering and Construction Contract: A user's guide. J Broome.
    ISBN 07277 2750 8
NEC and Partnering: The guide to building winning teams. J Bennett, A Baird.
    ISBN 07277 2955 1
ICE Conditions of Contract, 7th edition. Institution of Civil Engineers. ISBN 07277 3044 4
The New Engineering Contract: A legal commentary. A McInnis. ISBN 07277 2961 6

A catalogue record for this book is available from the British Library

ISBN: 978-0-7277-3115-9

Printed and bound in Great Britain by Bell & Bain Ltd, Glasgow

Contents

Preface

This book sets out to compare the increasingly popular New Engineering Contract (NEC) family of contracts with other construction industry standard forms. The absence of any litigation on the NEC family of contracts in the decade they have now been in use must be regarded as evidence of the success of their less adversarial approach, given the wealth of disputes under other standard forms, which have come before the courts in a similar time period.

The comparison aims to assist all levels of professionals involved in procurement in the construction industry to make informed choices and give balanced advice, as well as to evaluate advice given by others and to assess risk. There are chapters covering professional services contracts and construction contracts, as well as contractual arrangements for partnering. Building and engineering sectors of the construction industry are covered, as well as the possibility of procuring projects outside the UK.

This book will be invaluable to consultants, clients and contractors in the construction industry; it will also be a valuable tool for inquisitive students preparing for their professional exams. The book is not encyclopaedic, but rather concentrates on key comparative and contrasting issues, which can be examined further with reference to the contracts themselves.

Frances Forward

Contributors

The authors are all highly experienced in their respective fields within the construction industry and have been chosen for their extensive knowledge of both the NEC family of contracts and their sector-specific 'conventional' standard forms of contract.

Ernie Bayton FRICS FCIOB
Ernie Bayton works principally as a freelance chartered quantity surveyor advising employers, consultants, contractors and subcontractors on contractual and related issues. For 15 years he has undertaken public and in-house courses on all contract forms, including the NEC, for Thomas Telford Training as well as a number of other leading organizations, speaking mainly in the UK and occasionally overseas. He is a member of the NEC Users' Group and is an occasional article writer. Part of his time is still spent with Carillion Construction (formerly Tarmac) with whom he has worked for 33 years in a variety of senior roles including Chief Surveyor for Scotland.

Frances Forward BA(Hons) Dip Arch MSc(Const Law) RIBA FCIArb
Frances Forward is an architect, adjudicator and lecturer with 20 years' design, management and consulting experience in the construction industry, both in the UK and in Germany. Frances was a member of the working group that drafted the NEC Partnering Option X12 and she is also a member of the NEC Panel. Frances runs her own architectural practice, Forward Consult Ltd, and is an active user of the NEC family of contracts.

Richard Honey BSc(Hons) PGDip(Law) MSc(Lond) MRICS MCIArb
Richard Honey is a chartered quantity surveyor with Cyril Sweett Consulting in London, who also has a postgraduate diploma in law and the King's College MSc in construction law and arbitration. He is a member of the NEC Users' Group and has advised on the NEC for employers, contractors and consultants.

Roger Lewendon CEng FICE MIHT MCIArb MAPM
Roger Lewendon is a consulting engineer with over 40 years' experience in the UK construction industry and has contract experience gained in the construction of large infrastructure projects over many years. Roger is an active user of the NEC and provides consultancy advice and support to the organizations making the change. He is a member of the NEC Users' Group and also a member of one of the working groups involved in the future edition of NEC. He is also an active adjudicator and currently holds appointments on several major projects. He lectures widely on contractual matters throughout the UK.

Nigel Shaw FCIPS CEng MIMechE MInstD

Nigel Shaw is a procurement specialist with extensive experience working within leading construction client organizations both in the UK and internationally. Nigel was a member of the working group that drafted the Engineering and Construction Short Contract and is also a member of the NEC Panel. After many years with the electricity industry, since 1994 Nigel has been an independent procurement consultant, and has 'hands on' experience of developing and managing NEC contracts.

Brian Totterdill BSc(Hons) CEng FICE FIStructE FIPENZ FCIArb MAE FFB

Brian Totterdill is a chartered civil and structural engineer and chartered arbitrator with over 30 years' experience working on design and construction projects in the UK and overseas. Brian now concentrates on contractual problems, dispute resolution and the presentation of training courses. He has published several books on adjudication and dispute resolution and is the author of the recently published *FIDIC User's Guide – A Practical Guide to the 1999 Red Book*.

TW Weddell BSc DIC CEng FICE FIStructE ACIArb

Bill Weddell has spent most of his career as a consulting engineer. He was a member of the drafting team that produced the original NEC and has since continued as a member of the NEC Panel. He is a past member of the ICE Council and currently a member of the ICE Conciliation and Adjudication Advisory Panel. He lectures widely and runs training courses on the NEC Contracts as well as contract matters generally.

Introduction

Construction is by nature a relatively complex process and has long demanded contracts of greater sophistication than many other commercial transactions. The phenomenon of standard forms of contract plays a dominant legal and management role in contracting methods within the construction industry, first in relation to appointment of consultant design teams and secondly in relation to procuring both the construction and increasingly, the detailed design of projects by contractors.

It takes many years of practical experience and a keen interest in construction contracts to be in a position to give sound procurement advice, which genuinely considers all the options. The introduction of standard forms of contract into the construction industry was originally intended to improve fairness and to standardize rights and obligations. However, two particular difficulties are now apparent. First, the number of standard forms has proliferated. Secondly, sufficient dissatisfaction exists with many of those put forward, that a need is often perceived either to amend such forms heavily to suit individual projects, or indeed to draft entirely bespoke contracts in lieu.

Standard forms of contract are intended to operate on a 'consensus' basis and indeed the standard form drafting bodies are made up of a range of relevant and interested disciplines. It is, however, important to acknowledge that while they are inherently generic documents, they also rely on carefully considered insertion of project-specific parameters in order to safeguard optimum success.

A valid question is: how much dissatisfaction and bespoke drafting could be avoided if the most appropriate standard form were put forward in the first place, potentially saving time and money, as well as allowing greater certainty of outcome in terms of both performance and liability? Too often, apparently, professional advisers resort to the standard forms they are most familiar with and make them work as best they can, rather than considering all the available standard forms and putting forward the one most appropriate to the needs of individual teams and projects.

Construction projects have become more challenging, both technically and commercially. Many projects, whether public or private, require not only the combination of engineering and architectural skills, but also co-operation using specialist consultancy and industrial knowledge (already tangible in The Banwell Report, 1964). Accountability of employers and their professional advisers for effective procurement of construction projects has not diminished with consideration of the Latham and Egan Reports, but rather a climate has emerged in which the management of such accountability may become more sophisticated. In particular, risk management has become a key issue to be taken into account in choosing an appropriate procurement strategy.

The inspiration for this book stems from a desire to examine commonly used standard forms, old and new, across the construction industry and, in order to avoid random comparison, to benchmark all the others against the New Engineering Contract (NEC) family. The choice of the NEC as the 'control' family results from its breadth of application and therefore the ability to offer a comparative analysis with the entire sample. Whilst different chapters will inevitably appeal to members of different disciplines within the construction industry, an important intention behind such comparative analysis is to encourage greater interdisciplinary understanding and consequently better integration across multi-disciplinary teams. Above all, this book sets out to demonstrate that modern construction contracts can facilitate project management, as well as define legal relationships and that efficient procurement of complex projects demands a proactive, not a reactive, approach.

NEC compared and contrasted with JCT 98

Frances Forward

Introduction

The objectives of this chapter are first to compare the basis of the New Engineering Contract (NEC) and the Joint Contracts Tribunal (JCT) standard forms of construction contract in the context of building projects and secondly to analyse those aspects which display significant differences in their rationale and operation.

Type of contract

Conventional procurement analysis has long recognized that many building contracts provide the machinery to adequately control two out of the three parameters of time, cost and quality, but that an employer must often accept that the third parameter is somewhat compromised. Table 1.1 is an illustration of this view in relation to the three generic types of contract.

Table 1.1.

Type of contract	Traditional	Design and Build	Management
Parameters			
Time	—	✓	✓
Cost	✓	✓	—
Quality	✓	—	✓

Ideally, however, a balance between all three parameters of time, cost and quality is desired on individual projects. Moreover, a balance of the components of each parameter is also desirable: overall design and construction time *and* certainty of completion date; best value *and* cost certainty; specification quality *and* workmanship quality.

Form of contract

Standard forms of contract have conventionally been based on the principle that each 'type' of contract generates a new family member and that the actual form of contract is therefore chosen primarily as a result of time/ cost/quality analysis showing which of these parameters is least critical. The JCT documents are broadly founded on this premise, with a number of their standard form building contracts falling under the generic types listed in Table 1.1. NEC is, however, based on a new premise of flexibility, which allows the profile of particular projects to be matched in a much more sophisticated way. The detailed control of time, cost and quality is paramount and an inherent compromise of one of these parameters, as perceived with conventional 'traditional', 'design and build' and 'management' procurement, forms no part of the NEC philosophy.

Table 1.2 indicates the standard form building contracts that are under consideration.

Table 1.2.

Type of contract	Traditional	Design and Build	Management
Standard Forms			
JCT SF 98 (+CDPS)	✓	(with CDPS)	—
JCT IFC 98	✓	—	—
JCT MW 98	✓	—	—
JCT WCD 98	—	✓	—
JCT Management 98	—	—	✓
NEC 95	✓	✓	✓

Abbreviations: JCT SF 98: JCT Standard Form of Building Contract 1998
CDPS: JCT Contractor's Designed Portion Supplement 1998
JCT IFC 98: JCT Intermediate Form of Building Contract 1998
JCT MW 98: JCT Agreement for Minor Building Works 1998
JCT WCD 98: JCT Standard Form of Building Contract With Contractor's Design 1998
JCT Management 98: JCT Standard Form of Management Contract 1998
NEC 95: NEC Engineering and Construction Contract: 2nd Edition 1995.

Once employers and their consultant teams are familiar with NEC, it should be possible to gain a number of management and 'liability' advantages. The clarity and simplicity of the contract encourage both individual assessment of the contract strategy to suit a particular project and the subsequent implementation of standardized procedures. The contract documentation will be used actively as a management tool and decisions regarding design liability and risk allocation can be made on the basis of analysis of capability, rather than by following a 'traditional', 'design and build' or 'management' standard approach.

Contract amendments

While bespoke amendments to the JCT standard forms are often considered a solution to particular project requirements, it can be a very onerous solution. It may require considerable legal input to ensure consistent drafting of main and consequential amendments and it often unbalances the risk allocation between employer and contractor, with resulting cost implications and lack of legal certainty. NEC, however, with its flexible arrangement of core clauses, main option clauses and secondary option clauses, is designed to operate on a 'pick-and-mix' basis ensuring that particular project requirements are met through supplementary, not amended, clauses, capable of being chosen without the need for legal advice.

Contract preparation

Familiarity with older versions of the JCT family can lead to unsuitable preparation of the JCT 98 contract documentation, failing to take advantage of some of the newer provisions. NEC requires rigorous preparation of the contract documentation, in view of the project-specific emphasis and the consequential importance of the Contract Data and the Works Information and Site Information. The transparency of NEC means there is 'nowhere to hide' if any of the contract documentation is missing or unclear.

Design responsibility

Virtually all projects require a form of contract that allows the legitimate placing of design responsibility onto specialist contractors for a number of technical elements, for example mechanical and electrical detailed design. There are, however, relatively few standard forms that allow both such placing of design responsibility and have the contractual machinery in place to control adequately all three procurement parameters of time, cost and quality. Table 1.3 indicates the possible allocation of detailed design responsibility in the standard form building contracts that are under consideration; on this basis alone, some of the forms will not cater adequately for all the requirements of a particular project.

Table 1.3.

Design responsibility	Consultant	Hybrid	Contractor
Standard Forms			
JCT SF 98	✓	(CDPS or Nomination)	—
JCT IFC 98	✓	(Named Subcontractors)	—
JCT MW 98	✓	—	—
JCT WCD 98	—	✓	✓
JCT Management 98	—	✓	—
NEC 95	✓	✓ (Core Clauses 21 and 26)	✓ (Core Clauses 21 and 26)

Design liability

The standard of care for contractor design in those JCT contracts which envisage contractor design responsibility is 'reasonable skill and care', not 'fitness for purpose', albeit this is expressed indirectly by comparison with the standard of care of an architect or other appropriate professional designer. In contrast, the standard of care for contractor design in NEC is construed in relation to the Works Information, the default being closer to 'fitness for purpose', which can then be expressly limited to 'reasonable skill and care' by incorporation of secondary option M. The NEC approach reflects the desire to give flexible options that can suit specific requirements and, particularly, differing conventions within different countries and jurisdictions.

Contract machinery

The different genesis of the JCT and NEC contracts is clearly highly significant. The origins of the former go back more than a century, with its complex and subjective 'Victorian' drafting style, whereas the latter is barely a decade old, with its simple and objective 'modern' drafting style. There are many 'concepts' within the two forms which, while conceived wholly differently, nevertheless broadly equate; it is where the contract machinery to control such concepts is fundamentally different that interesting comparisons can be made.

Project management

JCT contracts have many instances of implied project management actions, but remain essentially a record of the parties' agreed rights and obligations. NEC embraces project management actions in an integrated manner, such that the legal rights and obligations are expressed in the context of specific communication types, to specific timescales. NEC requires contemporaneous conclusion of all aspects of such communications, which in turn requires adequate management resourcing from both Contractor and Employer for the duration of the contract. JCT allows sequential actions, often resulting in protracted conclusion to instructions, for example, time and cost implications of variations are often only partially assessed in subsequent valuation/certification, with detailed negotiations at final account stage. Both approaches tend to result in similar management resource requirements overall; however, NEC requires it evenly throughout the contract period, effectively achieving a running final account, whereas JCT tends to require an increased input at completion. A practical effect of this difference in approach is that JCT contracts tend to 'get left in the drawer' unless and until there is a dispute, whereas NEC contracts are much more likely to 'sit on the desk' for active reference as a project management tool.

Pricing mechanisms/payment

JCT sees the choice of pricing mechanism as an integral part of the form of contract and therefore only limited permutations are available. JCT SF 98 has introduced the option of activity schedules in lieu of bills of quantity, although their use would seem set to be very limited while it remains unclear on what basis either partially completed activities or variations for work not previously envisaged are to be valued. NEC, in contrast, envisages the pricing mechanism as a flexible component, to be chosen as a main option to supplement the core clauses. NEC further uses the pricing mechanisms to incentivize performance, for example, main option A requires activity schedule items to be 100% complete for interim payment, encouraging the Contractor to stay on programme in order to maintain cash flow. Main option C creates a target contract with 'pain or gain' for prices above or below the target cost being shared between Contractor and Employer in agreed percentages.

People

JCT contracts expressly refer to the Architect as contract administrator in traditional and management forms and assume the Architect is also the designer; in addition, they give an important role in relation to costs to the quantity surveyor. The JCT With Contractor's Design form provides for an employer's agent, who may or may not be the Architect, depending on whether 'novation' is favoured by the Employer. NEC gives the contract administration function to the Project Manager in relation to control of time and cost and to the Supervisor in relation to control of quality, including both workmanship and compliance with the Works Information. NEC therefore permits total flexibility as to whether the roles of Project Manager, Supervisor and designer are performed by one, two or three organizations. In practice, an architect can choose whether to offer to perform all three roles, if this is considered most appropriate to the procurement strategy for a particular project.

Programme

JCT requires the Contractor to produce a master programme; it does not, however, require the Contractor to prepare it in a particular form or to keep it up to date, nor is it conferred the status of a 'binding' contract document. This can impose onerous, if not virtually impossible, obligations on the contract administrator in the context of assessing extensions of time [see Henry Boot v Malmaison (2000)]. The new JCT Information Release Schedule from consultants to Contractor is not mandatory, indeed its use is apparently somewhat mistrusted, rather than being seen as a potentially useful control document. NEC, in contrast, makes the programme an integral contract document, which is kept up to date and can be used as a project management tool, both for monitoring progress accurately and in relation to incentivization. The NEC programme must contain not only details of construction sequence and information release, but also time risk and float allowances, giving a true picture of the critical path of the project.

Change control

The power to order change is seen by most employers in the construction industry to be an essential provision in their contracts. Many of the recommendations of the Latham Report can broadly be seen to relate to better control of that process of change. Both NEC and the 1998 revisions to the JCT family have procedures designed to manage change; the so-called compensation event procedure in NEC and variation provisions in JCT. An important difference between the contracts is that while JCT now attempts to assess the cost and time implications of a change early, it is not mandatory and review at final account stage has apparently not been unequivocally ruled out. In contrast, NEC makes contemporaneous assessment of cost and time implications of all changes both mandatory and binding, with no provision for later review.

Risk allocation

JCT contracts effectively have fixed risk allocation, which can be fine-tuned by the project-specific entries in the contract appendix, but not fundamentally altered, without actually amending the contract conditions. NEC allows for much more tailored allocation of risks to the party best able to carry them, or if appropriate, to specialist insurers. This is achieved principally by careful choice of the main option and any secondary options required, as well as by virtue of the detailed nature of the project-specific Contract Data. There should be no need to consider amending the core clause contract conditions in order to achieve the required risk allocation on a particular project.

Defects

JCT and NEC contracts both deal with the eventuality of 'patent' defects being identified before completion and 'latent' defects after completion. JCT contracts assume defects will be found by the contract administrator and rectified by instruction to remove or modify 'work not in accordance with the contract' before completion and remedied 'within a reasonable time' of notification by the contract administrator after completion and during the defects liability period. NEC contracts give a reciprocal obligation to Contractor and Supervisor alike to notify each other of defects found either before or after completion. The NEC defects correction period is a specific time period stated in the Contract Data, within which the Contractor is to correct notified defects after completion, up to the defects date, which is also stated in the Contract Data and equates with the end of the JCT 'defects liability period'. Defects include work not in accordance with the Works Information and the Contractor has an overriding duty to correct defects without specific instruction.

Completion

JCT contracts require certification of 'practical completion', and most envisage possible 'partial possession' and 'use or occupation', as well as enabling 'sectional completion', provided that a 'sectional completion supplement' has been incorporated to modify the contract conditions appropriately. NEC contracts allow for certification of 'completion' when the works are complete, or as complete as expressly required in the Works Information; there is also provision for certification of 'takeover' of parts of the contract works prior to completion. Sectional completion can be incorporated into NEC simply by adding secondary option L. JCT contracts require the issue of a 'final certificate' when all extensions of time, payments and defects have been finally concluded. NEC does not need a similar provision, due to the requirement to conclude all time, cost and quality issues contemporaneously.

Dispute avoidance and dispute resolution

JCT contracts comply with the statutory adjudication requirements of the Housing Grants, Construction and Regeneration Act 1996 (HGCR Act) and now offer a choice of tribunal, as between arbitration and litigation. The JCT family makes no provision for any form of alternative dispute resolution and no provision for early resolution of disputes on contracts that do not fall under the HGCR Act, i.e. owner occupier, or private finance initiative (PFI). JCT contracts remain essentially adversarial and currently contain no provisions for partnering arrangements. NEC was founded on the fundamentally non-adversarial principle of the parties acting 'in a spirit of mutual trust and co-operation' and on the requirement for the parties to give reciprocal 'early warning' (core clause 16) of matters which could prejudice time, cost or quality parameters, thus enshrining the first ingredients of partnering. NEC also provided for an interim dispute resolution process, 'adjudication', prior to this title acquiring statutory connotations under the HGCR Act; this remains intact for domestic projects not under the HGCR Act and for international projects, with separate provisions for those UK projects that do fall under the HGCR Act (core clause section 9 and secondary option Y(UK)2).

Application

JCT contracts are intended for 'domestic' projects, i.e. essentially in England and Wales only, albeit with related conditions for use in Scotland. NEC, in contrast, is designed to be operable internationally and there are useful provisions for international employers, for example the option of multiple currencies (secondary option K) and the ability to choose the law of the contract (Contract Data Part One).

Table 1.4.

Form of contract	JCT SF 98	JCT WCD 98	JCT MAN 98	NEC 95
Standard Provisions				
Contractor design	CDPS or Nomination	Completing detailed design	Works Contractors specialist design	Variable 0–100%, Core Clause 21/26
'Binding' contract programme	No	No	No	Core Clause 31
Acceleration provision	No	No	Possible (cl. 3.6)	Core Clause 36
100% Activity payment incentive	No – not 100% of Activity	Stages possible (Alternative A)	No – not 100% of Works Contracts	Main Option A
Bonus for early completion	No	No	No	Secondary Option Q
Binding time and cost, quotations for change	Optional (cl. 13A)	Optional (cl.12.4)	Optional (cl. 3.14 of Works Contract)	Compensation event procedure
Additional insurance requirements, e.g. PI	No	No	No	Contract Data Part One
'Independent' contract administrator	Architect	No	Architect	Project Manager (PM) acting for Employer
Direct Employer contract administration	No	Employer's Agent	No	NEC Short Contract 1999
Separation of administration and quality functions	Clerk of works under Architect's powers only	No	No	Supervisor with separate powers from PM
Provision for specific partnering arrangements	No	No	No	Secondary Option X12

Overview

A summary of how the NEC can not only offer much more detailed control of the time, cost and quality parameters simultaneously, but also give better overall project management control is shown in Table 1.4.

Conclusion

There is clearly no such thing as a perfect standard form of building contract for a particular project, but this knowledge should be balanced against the experience with entirely bespoke forms of building contract not generally being favourable. It is, therefore, proposed that a standard form of building contract should be chosen which has the machinery that matches the project profile most closely.

The key issues for most twenty-first century construction projects would seem to be:

- the need to control quality as well as maintain accountability for time and cost
- the need to allow for unforeseeable variables after execution of the contract, without poorly quantified time or cost implications
- the need to monitor the progress of the project accurately throughout.

The use of the NEC standard form of contract responds to these requirements, particularly by virtue of the 'pick-and-mix' contract strategy, the communication requirements and the compensation event procedure. The ability to control the time, cost and quality parameters is arguably better overall and easier to balance, by largely divorcing these parameters from any preconceived notion of 'type of contract' as found in the JCT family.

The ability to use the NEC Professional Services Contract for consultants' appointments (see Chapters 7–9), in parallel with the NEC Engineering and Construction Contract for the contractor's appointment and the NEC Engineering and Construction Subcontract for subcontractors' appointment allows an employer to benefit from the first truly back-to-back standard form documents across the construction industry.

NEC
compared
and
contrasted
with ICE

Roger Lewendon

Introduction

This chapter seeks to compare the philosophy of the 7th Edition of the Institution of Civil Engineers (ICE) Conditions of Contract, and its variants, with the New Engineering Contract (NEC) Engineering and Construction Contract, 2nd Edition 1995 and other documents in the NEC 'family'.

The ICE conditions

Traditionally, the procurement of civil engineering construction projects has used one of the editions of the ICE Conditions of Contract. These conditions were first issued in 1945, although their origins go back even further. To their considerable advantage they are legally tried and tested, and a custom of usage has developed by their extensive and successful use over the years.

The ICE Conditions of Contract presently published by the Institution of Civil Engineers consist of the 5th Edition (first published in 1973), the 6th Edition (first published in 1991) and the 7th Edition (Measurement Version) (published in 1999). In their published form, the 5th and 6th Editions now require significant amendment to take into account changes in legislation since their publication. The use of 5th and 6th Editions has continued in parallel, with many employers also amending both standard forms to produce their own 'bespoke' versions. They have a reputation for dogged use by traditional employers with their use sometimes inappropriate to the project needs and the employer's desired outcome.

As part of this family of contracts, the 3rd Edition of the Conditions of Contract for Minor Works was published in April 2001 and the 2nd Edition Design and Construct Conditions of Contract were published in September 2001.

Where a project can be completely designed and detailed and a Bill of Quantities produced then these standard conditions are a very capable form to use. However, if there are uncertainties which might require change, then the relative inflexibility of this form of contract starts to make itself felt.

The ICE Design and Construct Conditions of Contract, 2nd Edition, published in September 2001, radically departs from the normal ICE Conditions of Contract by making the Contractor responsible for all aspects of design and construction, including any design originally provided by, or on behalf of, the Employer. The Form of Tender provides for payment on a lump sum basis, but other forms of payment may be used.

Subcontracting is carried out under the complementary Civil Engineering Contractors Association (CECA) Forms of Subcontract which were last revised and published in 1998, as well as a wide variety of company-bespoke forms.

NEC: Engineering and Construction Contract

The NEC has much more recent origins and probably more closely reflects current procurement routes and needs. The contract is not only for civil engineering construction, but is intended to be used for any construction or building work.

The NEC is a procedurally based contract, requiring the parties to take certain actions in certain circumstances. In many cases it also includes sanctions against a party not taking the required action, or not doing so in the contractually required timescales. The NEC procedures give the opportunity for both parties to jointly provide more robust control and achieve increased certainty of project cost outcome.

The NEC includes an early warning procedure requiring either party to notify the other if a circumstance arises which might have an effect on cost, time or quality. This positive management approach encourages co-operation and provides a good basis for the use of partnering arrangements, leading to a reduction in disputes.

All parties involved in the project delivery process can be employed under one of the 'family' of integrated NEC contract forms which allows the parties to benefit from back-to-back contractual arrangements.

Comparisons

Table 2.1 compares the available variants of the ICE conditions and NEC. The table illustrates the more comprehensive range of back-to-back options available in NEC with the series of separate documents in the ICE conditions.

Comparison of the two forms

In looking at the two forms in more detail, the following gives some comparison of the more significant and different approaches of the two forms of contract.

Clarity and readability

The ICE conditions are written in semi-legal English, whereas the NEC is written in plain, more readily understood, commercial English. This has caused some unease to new users, perhaps because of what people have become used to over a long period of time. However, the 7th Edition of the ICE conditions and the 2nd Edition of the Design and Construct Conditions have benefited from the increased use of plain English, albeit in a modest way.

Table 2.1. *Procurement options ICE/NEC*

Procurement option	ICE conditions	NEC[1]
Priced contract with activity schedule	No	Option A
Priced contract with bill of quantities	7th Edition 2001[2]	Option B
Target cost contract with activity schedule	No	Option C
Target cost with bill of quantities	No	Option D
Cost reimbursable contract	No	Option E
Management contract	No	Option F
Design/construct	Design and Construct 2001[3]	All options 0–100%
Professional services	No	Yes[4]
Adjudicator	Agreement[5]	Yes[6]
Subcontract	CECA 'Blue Book' 1998[7]	Yes,[8] also short-form subcontract[9]
Minor works	3rd Edition 2001[10]	Short form[11]

[1]NEC: Engineering and Construction Contract, 2nd Edition 1995.
[2]ICE Conditions of Contract Measurement Version, 7th Edition 1999.
[3]ICE Conditions of Contract Design and Construct, 2nd Edition 2001.
[4]NEC: The Professional Services Contract, 2nd Edition 1998.
[5]Included in ICE Adjudication Procedure 1997.
[6]NEC: The Adjudicator's Contract Second Edition 1998.
[7]Civil Engineering Contractors Association Forms of Subcontract 1998 (Versions for use with 5th, 6th and design and construct forms of ICE conditions).
[8]NEC: The Engineering and Construction Subcontract, 2nd Edition 1995.
[9]NEC: The Engineering and Construction Short Subcontract, 1st Edition 2001.
[10]ICE Conditions of Contract Minor Works, 3rd Edition 2001.
[11]NEC: The Engineering and Construction Short Contract, 1st Edition 1999.

Management philosophy

The NEC proactive management philosophy encourages co-operation and the early joint resolution of problems. The simple structure of the documents encourages the use of procedures which can be applied across all the available contract options. This compares with the generally reactive management approach in the ICE conditions with problems being dealt with in a less structured way. This does not reflect against the ICE conditions, merely that the required management procedures in NEC are more robust requiring people to interface on matters sooner.

Management of the contract

The role of the Engineer in the ICE conditions, with its potentially conflicting responsibilities, is not used in NEC and decisions and directions are dealt with directly by the Employer (through their Project Manager) and the Contractor. The Employer's Representative in the ICE Design and Construct form acts as the agent of the Employer to supervise the design and construction of the works.

The role of the NEC Supervisor is limited to ensuring completion of the construction of the works in accordance with the specified standards set down in the Works Information. The NEC Project Manager and Supervisor are separate employer appointments with separate roles and responsibilities identified in the contract.

Procurement strategy

The clarity and simplicity of the NEC contract, with the choice of one of six optional contract strategies, enables individual consideration of each project. This flexibility allows the procurement of projects to be matched in a commercially realistic way to the employer's required outcome.

The NEC has a complete range of payment options such as priced contracts, target and cost reimbursable contracts, a management contract and the use of bills of quantities and activity schedules. With the six main options and a range of secondary option clauses each individual contract can be tailored from a set of standard clauses with little or no change to the standard documentation or the standardized procedures. The NEC contract procedures can be used actively as a management tool and decisions regarding design liability and risk can be allocated to the most appropriate party.

The ICE conditions has always been an admeasurement form and, although the current 7th Edition is a measurement version, it has been stated that the ICE documents issued since 1999 will be part of a 'family' of documents but the exact range of that 'family' has yet to be confirmed. Presently there are no target cost, cost reimbursable or management contract options. The use of the current standard form provides limited opportunity to overlap the design and construction processes, with the time-saving advantages that can arise.

The use of a bill of quantities is also increasingly being questioned and more employers are seeing advantages in the use of priced activity schedules available with NEC, often for no other reason than the simpler administration which results. This does lead to an increase in work and cost to the Contractor for the preparation of the priced activity schedule. The use of activity schedules can be considered advantageous to the Employer, as the risk of accuracy of the quantities of work is transferred from the Employer to the Contractor. It is essential, however, that there is a clear statement of the required scope of works in order that the activity schedule covers the full extent of the work required.

Payment options

The present 7th Edition of the ICE conditions is a measurement version of the contract. The 2nd Edition of the ICE Design and Construct Conditions allows for payment on a lump sum basis, although other forms of payment may be used. It also includes a pay/gain arrangement where the Contractor proposes a change to the employer's requirements and the parties share in the time and money consequences of the change. This can provide an incentive for the Contractor to seek potential improvements or cost savings.

The NEC offers a wider range of payment options with differing financial risk considerations and incentives: see Table 2.2.

Preparation of NEC tender enquiry documents

The NEC does, however, require more rigorous preparation of tender enquiry documents since no aspect of change to the stated scope of works is the responsibility of the Contractor. The Employer is required to provide the necessary information setting out the project requirements in the Works Information included in the Contract Data. The Works Information is, therefore, the ruling document clarifying the Contractor's responsibilities and ensures that he is in a position to tender with more confidence and provide a more realistic price to the Employer.

Subcontractors

The consideration of subcontractors is dealt with differently in the two forms, and is also different within the ICE conditions: see Table 2.3.

Under NEC there is no requirement, however, for a subcontract to be the same NEC main option as the main contract with the Employer. The benefit lies in the back-to-back arrangements which result.

Contractor's design

In order to optimize the design and eliminate unnecessary cost, flexibility in the design arrangements are often necessary. Increasingly employers are using contractor designs of the permanent work on the basis that contractors will use their expertise and consider 'buildability' in the design, with the result that the design is both safer and cheaper to build. The question of design liability is summarized in Table 2.4.

Table 2.2.

NEC option	Incentives	Financial risks	Other risks
Option A – priced contract with activity schedule	Payment on completion of activity encourages progress	With Contractor to complete the works within the tendered price	Contractor usually bears the risk of the completeness and accuracy of the activity schedule
Option B – priced contract with bill of quantities	Contractor motivated to keep within the tendered prices	With Contractor to complete the works within the tendered prices	Employer usually bears the risk of the completeness and accuracy of the bill of quantities
Option C – target cost with activity schedule	Shared financial incentive encourages co-operation to reduce cost	Shared on a pay/gain basis	Contractor bears the risk of the completeness and accuracy of the activity schedule
Option D – target cost with bill of quantities	Shared financial incentive encourages co-operation to reduce cost	Shared on a pay/gain basis	Employer usually bears the risk of the completeness and accuracy of the bill of quantities
Option E – cost reimbursable	No real incentive	Employer	Project out-turn cost uncertain
Option F – management contract	No real incentive	Employer	Project out-turn cost uncertain

Table 2.3. *Subcontracting*

	ICE 7th Edition	ICE Design and Construct 2nd Edition	NEC
Acceptance by Project Manager or approval by Engineer	No – notification only required. Engineer's representative can object	No – notification only required. Engineer's representative can object	Yes – Project Manager can refuse acceptance for stated reasons
Submission of subcontract conditions to Project Manager	No	No	Yes – unless ECS or PSC are used or Project Manager agrees that no submission is required
Contractor responsible for subcontractor's performance	Yes – except for nominated subcontractor	Yes	Yes
Nominated subcontractors	Yes	No	No

Table 2.4. *Design liability*

	ICE 7th Edition	ICE Design and Construct 2nd Edition	NEC
Amount of contractor design	Stated elements only	Envisages complete design	Any level from 0 to 100%
Contractor design liability	No liability for design of permanent works	Reasonable skill care and diligence	Fitness for purpose
Change to Contractor design liability	By express provision	No	By inclusion of option M liability can be changed to 'reasonable skill and care'
Responsibility for employer's design	Employer	Contractor	Employer
Designer's appointment	No conditions stated	No conditions stated	Anticipates the use of PSC

Programme and time

The philosophy in relation to a formalized construction programme between the ICE and NEC forms is quite different.

The ICE conditions require the production of a construction programme and its updating, if it becomes unrepresentative of actual progress. Traditionally, the contract programme often becomes a vehicle for the consideration of delay and the justification for extending the time for completion.

NEC, however, envisages the programme being used by all parties to consider progress, delays, delays to completion and to enable the regulation of future actions to achieve an acceptable outcome. It can be particularly helpful in looking at alternative scenarios for carrying out additional work or dealing with delays and assist in providing the optimum solution against the project objectives.

Early warning procedure

This is seen by many as the 'jewel in the crown' of NEC. The contract places an obligation on the parties to each notify the other of any matter which could increase the prices, delay completion or impair the performance of the works in use. The notifying party can also call a meeting to discuss the matter and the other party is obliged to attend.

The importance of this requirement is that the parties are motivated to identify problems as early as possible and have a proactive approach to find a solution jointly, rather than putting off decisions or ignoring their resolution. The procedure covers any matter and has no consideration as to whose fault it was, or where the liability lies. The desired outcome is to provide agreed solutions to problems before they have any adverse effect on progress.

Valuation of change and compensation events

There is a fundamental difference between the two forms of contract in the valuation of change.

Under the ICE conditions, the change is usually instructed by the Engineer and the price is subsequently determined, based on the tender pricing structure in the contract. The pricing of change is often carried out some considerable time after the work is carried out, creating uncertainty in the eventual out-turn cost to the Employer and delay costs are usually dealt with as a separate issue. The ICE conditions 7th Edition does, however, include a provision for the Engineer to request an all-inclusive quotation from the Contractor, with the intention of agreeing this before work is ordered or starts.

NEC includes the concept of 'compensation events'. These are events outside of contractors' control and for which they are entitled to be compensated. The contract defines what are compensation events and includes instructed change to the Works Information (i.e. a 'variation'). The Contractor submits a quotation for the changes to both time and cost based on actual cost or forecast actual cost as defined in the contract. The Project Manager's acceptance of that quotation, or his or her own assessment, implements the change. The intention is that the cost and time implications are established before the implementation of the change. The NEC approach has the advantage that the Project Manager knows what costs the Employer is committed to, usually before the work is started.

Defects

In the ICE conditions there is no clear definition of 'defect', except by a very general subjective description. To become entitled to the Defects Correction Certificate (i.e. 'at the end of the maintenance period') the Contractor is required to rectify all defects and complete all outstanding work to the Engineer's satisfaction.

The NEC approach differs in that a 'defect' is a defined term and the Contractor and the Supervisor are both required to notify the other of defects they find. NEC also has a 'defects date' and a 'defects correction period'. The 'defects date' is the date at which the work can be considered to be free of defects and is the equivalent of the traditional 'maintenance period'. The 'defects correction period' is the much shorter period after completion in which the Contractor is required to correct notified defects.

Dispute resolution

The ICE Conditions of Contract offer conciliation and arbitration as dispute resolution procedures together with adjudication in accordance with the Housing Grants, Construction and Regeneration Act 1996 (HGCR Act).

NEC requires that a dispute is first referred to the adjudicator and if this fails then it is referred to a further tribunal. A choice of tribunal is offered which, in effect, means arbitration or litigation. Adjudication in the UK is a revised procedure in accordance with the HGCR Act. For contracts outside of the UK, or outside the coverage of the Act, the standard adjudication provisions included in the contract can be used.

Conclusions

A construction contract is basically no different to any other commercial arrangement which requires the delivery of a specific outcome to a specified quality, at an acceptable cost and in an acceptable time. As well as the need to control quality, there is also the need to monitor the progress of the project realistically and control out-turn cost, particularly where there are financial constraints or limitations. Certainty of outcome is important if employers' capital investments are to perform as intended in terms of their businesses and that value for money is achieved.

The NEC standard form of contract is able to respond to these requirements by virtue of the flexible contract strategy that enables the optimum procurement strategy to be used, based on the employer's objectives. Each individual contract can be tailored from a set of standard clauses, rather than the changing of standard clauses on a 'bespoke' basis. The early warning and compensation event procedures encourage good proactive management and the increased and improved ability to control time, cost and quality should, therefore, become a realistic expectation with benefits to all parties to the contract.

The ICE conditions have been used very successfully to deliver a large quantity of work over many years and they are legally tried and tested. There have been problems, not least the adversarial and claims culture which has developed. The present development of the conditions appears to be addressing some of these factors, albeit in a modest way. The inclusion of the facility for a prior quotation in the 7th Edition conditions and an early warning procedure in the current design and construct form are particular examples.

NEC compared and contrasted with FIDIC

Brian Totterdill

Introduction

This chapter compares the New Engineering Contract (NEC) with the conditions of contract that are most commonly used for international projects. The conditions of contract which are published by the Fédération Internationale des Ingénieurs Conseils (FIDIC) are intended to be used for international projects and are only rarely used for projects in the UK. By contrast, the NEC family of contracts are used primarily for projects in the UK, but are also intended to be suitable for use in other countries.

This chapter compares the NEC Engineering and Construction Contract, 2nd Edition 1995 (NEC), with the FIDIC Conditions of Contract for Construction, 1st Edition 1999 (FIDIC).

Contract philosophy

The guidance notes to the NEC include a review of the background and objectives of the NEC, including the statement that 'A fundamental objective of the [contract] is that its use should minimise the incidence of disputes'.

The FIDIC contracts appear to have been written on the principle that, in construction, there will always be problems and a potential of disputes. One of the roles of the conditions of contract is therefore to provide a framework that will help identify and resolve problems or, if the problem develops into a dispute, will ensure that accurate evidence is recorded. This will assist the tribunal; both the immediate dispute adjudication board and finally an arbitration tribunal. The project management procedures have therefore become more complex in the 1999 contracts. In particular, the procedures have been developed for the Contractor to submit monthly reports and to notify and provide information in support of any claims.

The layout and style of writing

When the NEC is compared to the traditional FIDIC contracts, the layout of the NEC is much easier to follow. The 1999 FIDIC contracts have clearly benefited from this comparison and the layout and style have been considerably improved.

The NEC uses the present tense throughout whereas FIDIC uses the traditional 'shall do' terminology.

The family of contracts

Both the NEC and FIDIC have published families of contract documents. Comparing the two families, the NEC is presently the more complete bundle. The NEC includes provision for both employer design and contractor design in the same contract and can be used for electrical and mechanical engineering as well as for civil engineering and building work. The NEC family includes a partnering option, a professional services contract, a subcontract, a short contract and an adjudicator's contract. The professional services contract can also be used when the Contractor appoints a consultant to carry out design work. Each of the NEC contracts is accompanied by flow charts and guidance notes, which include guidance on the preparation of tender documents.

The FIDIC family was revised and rearranged with the publication of the 1999 contracts. The conditions of contract for construction are for use when the design is provided by the Employer, with a separate contract for use when the design is to be provided by the Contractor. However, in practice most contracts require some design work to be carried out by the Contractor and so some design provisions are required in most construction contracts. FIDIC have also published a contract for use when the Contractor is responsible for the complete Engineering Procurement and Construction (EPC) Turnkey projects, and a short contract. Some guidance notes and the dispute adjudication board agreements are included in the same document as the conditions of contract. The FIDIC contract guide includes a detailed commentary on each clause. The FIDIC consultant's agreement, 3rd Edition published in 1998, has not been revised. The FIDIC tendering procedure was published as a separate document in 1994, but this has been effectively superseded by tender procedures that are included in the contracts guide. However, the subcontract, which was also published in 1994, requires extensive changes if it is to be used with the 1999 contracts.

The layout and alternatives

The layout of information within the contracts and the procedures for introducing alternative or additional clauses are completely different in the two contracts.

The NEC is based on core clauses, which are in nine sections, together with a selection of options. The guidance notes include a warning that: 'Additional conditions should be used only when absolutely necessary to accommodate special needs such as those peculiar to the country in which the work is to be done.'

The FIDIC layout is based on general conditions and particular conditions. The general conditions are intended to be incorporated into every contract. Any changes or additional clauses are included in a separate document as particular conditions. The FIDIC publication includes guidance and examples of a wide range of particular conditions that may be used to suit the requirements of the Employer for the particular country and project. Additional clauses can be incorporated as required, but care must be taken to ensure that all clauses are properly co-ordinated with the general conditions.

The Contract Data

Both the contracts include provision for employers to provide basic information concerning the project and for contractors to add further information when submitting their tenders. The NEC refers to this document as the 'Contract Data', whereas the FIDIC document is shorter and is called the 'appendix to tender'.

The NEC Contract Data are more comprehensive and include reference to the documents that contain the Works Information, the Site Information, the Works Information for any contractor design and the bills of quantities if required. The documents that form the contract documents are listed in the sample form of agreement.

FIDIC includes provision for a list of documents in the contract agreement. The conditions of contract also refer to other information that is included in the particular conditions or in other documents.

The project management organization

Both contracts have moved away from the traditional organization under which an engineer or architect is appointed by the Employer, but acts as an impartial professional when making decisions on matters such as payments, claims and extensions to the contract period. Both contracts provide for the initial decision on these matters to be made by a representative of the Employer, acting within certain constraints, whose decision is subject to review under an adjudication procedure during the construction period.

The NEC contract incorporates a Project Manager, who is appointed by the Employer and who manages the contract for the Employer, with the specific intention of achieving the employer's objectives for the completed project. The project manager has considerable authority and the basis of his or her actions is determined in the contract. Where the Project Manager makes a decision, the contract generally requires the possible reasons to be given why the Project Manager may reject some request from the Contractor. In addition, the NEC has provision for a Supervisor, who is also appointed by the Employer. The actions and decisions that can be taken by the Supervisor are stated in the contract, but the role is to check that the works are constructed in accordance with the contract. The Project Manager and Supervisor may delegate any of these actions. Any action of the Project Manager or Supervisor that is disputed by the Contractor can be referred to the adjudicator.

The NEC, at sub-clause 10.1 is clear that: 'The Employer, the Contractor, the Project Manager and the Supervisor shall act as stated in this contract and *in a spirit of mutual trust and co-operation*' (my emphasis).

The 1999 FIDIC contracts retain the traditional position of the Engineer, but that role is clearly defined as being part of the employer's personnel. Matters such as payments, claims and extensions of time are decided by the Engineer. Interim payment certificates are for 'the amount which the Engineer fairly determines to be due', as sub-clause 14.6. Any claims are initially determined by the Engineer, as sub-clause 3.5. The Engineer is required to 'consult with each Party in an endeavour to reach agreement' and, failing agreement, to 'make a fair determination in accordance with the Contract, taking due regard of all relevant circumstances'. The emphasis throughout is for the Engineer to be fair and act in accordance with the contract. If either party is not satisfied it can refer the matter to the dispute adjudication board, as discussed later in this chapter.

Some important provisions in the conditions of contract

Both contracts are for construction projects, so they must include provisions to cover the same requirements and circumstances. A detailed clause by clause comparison would not be appropriate and would obscure the basic differences in principle. The following comparisons relate to some important provisions for which the contracts have a different approach.

Payment procedures

The NEC provides alternative options for different methods of payment such as a target contract or cost reimbursable contract and for the use of either a bill of quantities or an activity schedule. Each option provides a package of sub-clauses that are inserted into the core contract.

FIDIC is essentially a remeasured contract, based on a provisional bill of quantities. The method of measurement is required to be in accordance with the bill of quantities 'or other applicable Schedules'. Guidance is given for the preparation of a cost-plus or lump-sum contract but the actual alternative clauses must be written by employers to suit their requirements.

A comparison of the time periods from the date when an item of work is recorded to the payment by the Employer is given in Figure 3.1. Clearly, the NEC gives a quicker payment which benefits the Contractor. However, some international employers, who rely on payment from international finance institutions, might need to increase the NEC payment period.

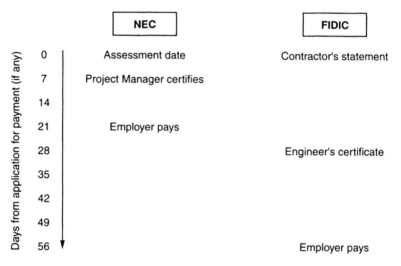

Figure 3.1. *Payment procedures.*

Variations

Any construction contract must include provision for employers, or their representatives, to issue instructions for changes to the work that was included in the original tender.

The essential differences between the NEC and FIDIC procedures are the procedures before issuing the instruction and in the valuation of the change. Under the NEC the Contractor submits a quotation, or alternative quotations, for the proposed changes. If the Project Manager accepts the quotation then the Contractor is paid this quotation.

Under FIDIC, the Engineer issues an instruction and the work is valued, generally based on the original contract rates, after the work has been completed. FIDIC also includes provisions for the Engineer to request a proposal from the Contractor before issuing an instruction or for the Contractor to instigate a value engineering procedure. These procedures are a matter of proposals and are not fixed price quotations as under the NEC.

Unexpected situations

The NEC includes provision for compensation events, which is a procedure for managing events that are at the Employer's risk. This is intended to enable the Project Manager and Contractor to manage the consequences of an unexpected situation in a positive way. Sub-clause 60.1 gives a list of events, including variations, which may give the Contractor an entitlement to additional time or payment. The procedure is the same as has been described for a variation, with the Contractor submitting a quotation for the time and cost. These situations are often uncertain and sub-clause 61.6 provides that, if the effects of a compensation event are uncertain then the Project Manager may state assumptions on which the quotation must be based.

The NEC includes a provision, at clause 16, for either the Project Manager or the Contractor to call an early warning meeting. This provision requires the Contractor and the Project Manager to co-operate in considering how to overcome any problem and to agree on the solution which is in the best interests of the project. If the Contractor fails to give the required early warning notice then the Project Manager may take into account any costs that would have been saved if the notice had been given.

The FIDIC procedure is based on written notices with the Contractor giving an advance notice of any potential claims situation, a further notice of any actual cost or delay and then providing detailed records of the consequences. The Engineer issues any necessary instructions and decides whether the Contractor is entitled to additional time or payment. The FIDIC requirements are scattered throughout the contract, but the provisions of clause 20 apply to all claims. This includes a provision that, if the Contractor fails to give the required notice within 28 days from becoming aware of the event, then the Employer will be discharged from all liability for the claim.

A comparison of the procedures that may be required when an unexpected situation arises is given in Figure 3.2.

Provisions for the resolution of disputes

Both contracts include provisions that if the Contractor is not satisfied with a decision by the employer's representative, the resulting dispute can be referred to immediate adjudication. Under the NEC the UK adjudication procedures are an alternative option so for international projects the adjudication procedures in the core clause would apply.

A comparison of the dispute resolution procedures is given in Table 3.1.

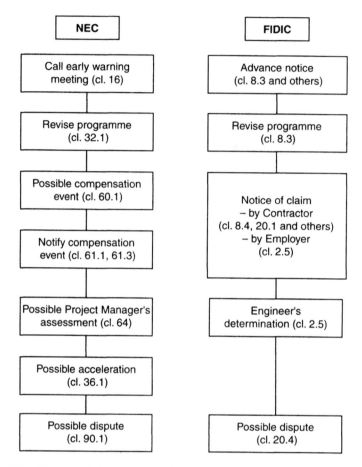

Figure 3.2. *Unexpected situation procedures.*

Conclusions

Any standard conditions of contract may need to be modified when used for an international project. A requirement for changes or additional provisions may result from a number of different causes, such as:

- a requirement of the applicable law
- a statutory regulation which applies to the project
- a requirement of the bank or finance authority who are funding the project
- a custom or procedure which is generally adopted in the country of the project
- a preference of the employer.

Table 3.1. *Adjudication and arbitration*

	NEC	FIDIC
Adjudication board	Single person	One or three people
Selection	Named in contract data or agreed	Agreed or independent nomination
Appointment	Start of contract	Start of contract
Starts work	When dispute arises	Regular site visits
Procedure	As stated in contract	As stated in contract
Decision	Enforceable under contract	Must be implemented under contract
If decision not accepted	Refer dispute to tribunal as stated in Contract Data, may be arbitration	Amicable settlement followed by international arbitration

Any proposed changes must be considered very carefully before incorp-orating additional clauses into the NEC options or FIDIC particular conditions. Any changes or additions, other than information for which there is provision in the Contract Data or appendix to tender, will probably affect other clauses or disrupt the contract procedures and so the implications must be studied in detail.

Before incorporating any change or addition to the standard contract it is essential to consider whether the proposed change or addition is essential. It may need to be considered whether it is already covered by a general requirement that the parties must abide by the applicable law and statutory regulations, or whether it is just a matter of choice between a personal preference and a recognized international contract procedure.

When either the NEC or FIDIC is used for an international project it must be considered as a whole, including the options or recommended particular conditions and in conjunction with the published guidance notes. It is essential to ensure that the change does not conflict with the philosophy or disrupt the procedures in the standard contract.

NEC compared and contrasted with GC/Works

Ernie Bayton

4

Introduction

The GC/Works family of contracts are UK government forms of contract for construction works. Presently published by the Property Advisers to the Civil Estate (PACE), GC/Works/1 to GC/Works/4 were heralded as being the first new contract conditions fully compliant with the Latham Report (*Constructing the Team*) and the Housing Grants, Construction and Regeneration Act (1996).

Since their publication, the initial range of contracts has been supplemented by further conditions (GC/Works/5 to GC/Works/10 and GW/S) serving other purposes. Comparing and contrasting GC/Works contracts with the New Engineering Contracts (NEC) reveals that while there are some areas of common ground, much clear water exists between them.

This chapter concentrates on GC/Works 1 to 4 but makes references where appropriate to GC/Works/5 to GC/Works/10 and the GC/Works Subcontracts.

Background

GC/Works contracts have a long tradition. The present range, launched in 1998, builds on a series of contracts devised originally before the second world war for government defence works. Each form targets a specific work type providing standard conditions for a specific procurement route, resulting in numerous stand-alone conditions of contract. By contrast the NEC group of documents are less than 10 years old. They require greater emphasis to be placed on choosing the correct procurement route before determining, from the available options, conditions that are tailored to the project.

Although devised for government contracts (not all government-sponsored work is executed under GC/Works conditions), private sector clients occasionally use GC/Works conditions. Additionally, the GC/Works contracts are targeted specifically towards building and civil engineering works, or mechanical and electrical engineering works. The NEC, not being aligned to any particular client user group, is suitable for public and private sector work and is multidisciplinary.

GC/Works contracts are designed for use in the UK whereas NEC contracts have international application.

Risk allocation

Risk allocation between employers and contractors under the new GC/Works contracts has changed substantially from its predecessors.

Originally perceived to be documents written by the Employer, for the Employer and with the Employer's interests at heart, recent versions of GC/Works contracts have moved towards a fairer allocation of risks between the Employer and Contractor. Part of that movement embodies Latham's principles of 'fair dealing and team working' which are incorporated in the conditions at Condition 1A. Failure to include the Project Manager and quantity surveyor within the expressed obligations of fairness, good faith and mutual co-operation at Condition 1A(1) is a disappointing omission. Additionally, Condition 1A's numbering has the unfortunate connotation of an afterthought, or an option capable of deletion, unlike the NEC's embodiment of mutual trust and co-operation established at the outset.

Publications

The GC/Works family of contracts include at least 19 different contract forms, GC/Works/1 to GC/Works/10 plus subcontracts, together with model forms and commentaries, providing conditions for consultancy and construction contracts. Despite their proliferation, GC/Works documents fall far short of the NEC's greater harmonization and range of options achieved with a smaller number of publications. The GC/Works range lacks many options available to NEC users, has no adjudicator's contract, partnering option or flow charts that accompany NEC guidance notes.

Format

GC/Works/1 to 4 documents derive from a common source: see Table 4.1.

Table 4.1.

GC/Works contract	Derived from	Notes
GC/Works/1: Part 1	Foundation document	With quantities
GC/Works/1: Part 2	GC/Works/1: Part 1	Varied for without quantities
GC/Works/1: Part 3	GC/Works/1: Part 1	Varied for design and build
GC/Works/2	GC/Works/1: Part 2	Simplified version
GC/Works/3	GC/Works/1: Part 1	Varied for mechanical and electrical works
GC/Works/4	GC/Works/2	Simplified version

GC/Works/1: Parts 1, 2 and 3 incorporate optional conditions for deletion when not applicable. Parts 2 and 3 use identical condition numbers to Part 1, but where inappropriate state 'not used'. NEC contracts adopt a simpler approach using core clauses common to all contracts with added clauses discrete to the main option and secondary options chosen.

Table 4.2.

Provision	GC/Works/1	GC/Works/3	NEC	Notes
Employer's agents	Project Manager, quantity surveyor, clerk of works and resident engineer	No quantity surveyor, otherwise as GC/Works/1	Project Manager and Supervisor	Employers may limit project manager's powers under GC/Works
Powers of delegation	By Project Manager and quantity surveyor	By Project Manager	By Project Manager and Supervisor	Project Manager has wider powers under NEC
Model forms	23 – use not mandatory	28 – use not mandatory	Sample tender and agreement forms only	NEC parties can use their preferred forms
Contract documents defined	Yes	Yes	Listed in Agreement	GC/Works contracts need amendments to incorporate other documents
Priority order for documents specified	Yes	Yes	No	NEC ambiguities or inaccuracies are resolved by Project Manager's instructions
Contractor's key people identified in contract	No	No	Yes	NEC allows replacement by agreement
Early warning procedures	Alluded to	Alluded to	Yes	NEC rules are direct and defined

Feature				Comments
Notifications to be separate from all other communications	No	No	Yes	Easier identification of notices under NEC
Use of time limits	43 Conditions impose 14 different time limits, all except extension of time provisions, may be extended by agreement prospectively or retrospectively	As GC/Works/1	Period for reply common to all communications (extended by agreement prospectively), except a few specific clauses adopting discrete timescales	NEC period for reply stated in Contract Data allowing it to differ between contracts
Progress meetings	Yes	Yes	None specified	NEC contracting parties free to agree
'Completion' and 'defect' defined	No	No	Yes	NEC definitions reduce subjectivity
Any employer's unwillingness to accept works before completion date stated	No	No	Yes	NEC specifies unwillingness at tender stage; Contractor can assess risk
Defects	Defects once remedied are further subject to the relevant maintenance period	As GC/Works/1	Defects correction period (DCP) attaches to each defect	Contractor pays amount assessed by Project Manager if defect is not remedied within its DCP

Administration

Essential to the success of any project is good administration. Table 4.2 compares some key administrative provisions between GC/Works/1 and 3 contracts with the NEC.

Design

Employers, when considering their procurement options, must decide where best to obtain the design expertise for their project, either in-house, external consultant, contractor or supplier. Involving contractors in the design process has rightly increased in recent years. GC/Works/1 to 4, GW/S Subcontracts and NEC construction contracts address client, contractor (or subcontractor) design involvement as summarized in Table 4.3.

Table 4.3.

Contract form	Design responsibility		
	Client	Partial contractor	Contractor
GC/Works/1: Part 1	Yes	Yes	No
GC/Works/1: Part 2	Yes	Yes	No
GC/Works/1: Part 3	No, except Employer require-ment input	No	Yes
GC/Works/2	Yes	No	No
GC/Works/3	Yes	Yes	Yes
GC/Works/4	Yes	No	No
GW/S Subcontracts	Yes	Yes	No
NEC/ECC	Yes	Yes	Yes
NEC/ECSC	Yes	Yes	Yes
NEC/ECS	Yes	Yes	Yes

All NEC construction contracts facilitate 0–100% contractor design involvement without the need for separate stand-alone conditions. Since designers for the Employer are to 'develop the design to the point where tenders for construction are to be invited', the Works Information establishes any contractor's design responsibility beyond the tender invitation stage.

Both groups of documents contain adequate procedural issues in relation to design, although under the GC/Works Forms certain issues may be subject to 'or Instructed by the Project Manager' potentially leaving the Contractor exposed.

GC/Works/5 consultancy agreements, like their construction contract counterparts, lack the flexibility and scope of the NEC Professional Services Contract (PSC). The PSC reflects the main option payment mechanisms of lump sum, re-measurement, target cost and cost plus available under other NEC contracts

Key differences between NEC and GC/Works 1 and 3 contracts *vis-à-vis* contractor's design provisions are given in Table 4.4.

Programmes

Modern contract conditions pay greater attention to programmes. They feature prominently in GC/Works/1, GC/Works/3 and NEC construction and consultancy contracts. GC/Works 2 and 4 are silent. NEC contracts are 'programme driven'. All parties are obliged to buy into its constraints including 'Others', for example, statutory authorities, direct contractors, etc. Under NEC contracts, accepted programmes, through the detailed revision procedures effectively become 'travelling documents' moving with the parties throughout the project, making it a more effective tool.

Table 4.5 analyses the provisions of GC/Works 1 and 3 and the NEC in relation to programmes.

Payment

Styles of contract

GC/Works/1 to 4 are all lump sum contracts, thus straight-jacketing them-selves, preventing the use of other payment mechanisms, and making the Contractor the main bearer of the financial risks. Contracts rely either on bills of quantities, schedules of rates or other pricing documents in an unspecified form. In contrast, NEC contracts, under their six main options, provide a wider choice of payment mechanisms allowing the financial risks to be allocated more appropriately between the parties on a discrete contract basis. In practical terms the six forms within GC/Works/1 to 4 only cover two of the six NEC main options.

NEC options B and D use bills of quantities. Options A and C use activity schedules, a pricing document produced by contractors (at their risk) at tender stage. Use of activity schedules is well developed within the NEC and facilitates tender comparisons and financial management, monitoring and control during the currency of the contract.

The range of payment-related options available under NEC contracts are compared with GC/Works/1 and 3 in Table 4.6.

Table 4.4.

Key difference	GC/Works/1 and 3	NEC/ECC	Comments
Document specifying contractor design role	Abstract of particulars together with other contract documents	Works Information	
Reasons stated in contract for Project Manager not accepting Contractor's design	No	Yes	Under NEC other reasons can create compensation event entitlements
'Fitness for purpose' liability	Yes – if alternative Condition 10B used	Yes – if secondary option M is not incorporated	NEC assumes fitness for purpose as a basic position
Reasonable, skill and care liability	Yes – if alternative Condition 10A used	Yes – if secondary option M is incorporated	Under NEC secondary option M the burden of proof rests with Contractor
Provisions capping a contractor's financial liability to Employer for defects	No	Yes	Under NEC strict tests apply, e.g. rules exclude patent defects
Professional indemnity insurance requirement	Yes – if Condition 8A applies	No	Additional insurance requirements can be included in NEC Contract Data
Interlocking professional services contract	No	Yes	The NEC/PSC is useable by employers, contractors or subcontractors for consultants
Approval of design from others	Not stated	Yes	NEC defines Others

Table 4.5.

Subject	GC/Works/1 and 3	NEC
Contract document	Yes	Incorporated by reference
Detailed content specified	Yes	Yes – better defined than GC/Works
Specified obligation to recognize planned completion, float and time risk allowances	No	Yes
Submit with tender	Yes	Optional
Acceptance procedures	No	Yes
Reasons specified for Project Manager not accepting programme	No	Yes
Programme to be based on contract period	Yes – a questionable obligation if the stage payment chart option does not apply	No – planned completion is recognized
Used for interim payment assessments	Yes – only where stage payment chart option applies	Yes – main option A only for completed activity or group of activities
Revision provisions	Only at contractor's initiative	At Project Manager or Contractor's initiative, or frequency specified in contract
Revised programme content specified	No specific requirement – other than the content for all programmes	Yes
Timetable for revised programme submissions	No	Yes
Timetable for revised programme acceptance	No	Yes
Payment remedy for contractor's failure to submit conforming first programme	No	Yes
Required for assessing delays	No	Yes – for quotation and assessment

Table 4.6.

Payment feature	GC/Works/1 and 3	NEC
Payment options	Lump sum contracts only	Wider range of options (A–F) including lump sum, re-measurement, target contract and cost reimbursable options provide greater flexibility
Extra features not common to both	More restrictive list of options including: • mandatory advance payment bond if advanced payment made • mandatory retention payment bond if retention not held	Wider range of other features including: • well-developed schedule of cost components • fairer finance/interest payment provisions • optional advanced payment bond • multiple currencies • price adjustment for inflation • low performance damages • trust fund

Interim payments

Interim payment provisions under GC/Works contracts vary. Many use stage payment charts introduced as a means of speeding up interim payment procedures. Their use has not been adopted as widely as the drafters intended. Additionally, milestone payment and 'traditional' valuation methods are available. One option is used per contract. Amounts payable for varied work are frequently valued differently to the chosen option for the original work, thus applying different payment regimes to different parts of the work, some of which are subject to retention and others not.

NEC payment entitlements vary according to the choice of main option. Once chosen, they remain constant throughout the contract.

Changes

Rules for valuing changes differ between the two forms. GC/Works/1 and 3 contracts employ a combination of:

- 'traditional' methods of valuation including prolongation and disruption
- lump sum quotations including prolongation and disruption
- prolongation and disruption (not included in either of the above).

In practice the rules are complex, generate different payment entitlements depending on whether they are agreed or not and have different time constraints applied to them.

NEC construction contracts use simpler rules under which compensation event procedures contemporaneously deal with time and money.

Final accounts

GC/Works/1 and 3 forms, which are more prescriptive than the NEC regarding final account procedures, include provisions for an agreement timetable, issuing draft accounts for agreement or disagreement (with reasons) and limited default mechanisms leading to a deemed acceptance. Unfortunately some of the procedures in the conditions are incomplete (e.g. actions by the parties following a legitimate objection by the Contractor to a draft final account). The NEC is less prescriptive, preferring to rely on the obligation of trust and co-operation, coupled with the obvious commercial pressures to obtain agreement and consequential settlement of outstanding sums becoming due.

Compensation events

Compensation events entitlements under the NEC contracts have significant advantages over equivalent GC/Works contracts. Key advantages are:

- compensation events are singly located
- time and money are considered together
- employers may insert additional compensation events at tender stage
- Project Manager may notify compensation event entitlements in certain circumstances
- quotations are essential for assessing entitlements
- quotations comprise time, money, supporting details and programmes (where remaining work is affected)
- main options vary the purpose of quotations
- timetable applies for submitting and accepting quotations
- quotations for proposals or alternatives may be provided
- reasons for Project Managers not accepting quotations are stated
- Project Managers make their own assessments in stated circumstances

- delays are assessed against planned completion in the accepted programme
- monetary entitlements are assessed as actual cost
- actual cost and disallowed cost are defined
- schedules of cost components (SCC) are provided
- an agreed fee percentage covers items not reimbursable under the SCC
- where the effects of a compensation event are too uncertain to be forecast reasonably, Project Managers provide assumptions for assessment purposes
- quotations based on Project Managers' assumptions are revisited if later proved to be incorrect
- decisions are final unless based on assumptions stated by the Project Manager
- facilitates better financial control of the project since post-contract claims and valuation disputes should be eliminated
- creates more certainty for all parties.

Subcontracting

Subcontracting and subcontracting methods continue to evolve. Contractors often sublet high proportions of their work. Principal contract conditions consequently must recognize practice in the workplace and ensure that adequate second-tier contract provisions are available.

GC/Works 1 to 4 limits its horizons, assuming a traditional approach to subcontracting. Standard subcontracts for use in conjunction with GC/Works/1 to 3 (GW/S) have been published by PACE. The Civil Engineering Contractors Association (CECA) have also published standard subcontracts for GC/Works/1: Parts 1 to 3. Neither group of publications adequately mirrors those of the principal contracts.

NEC construction subcontracts adopt a broader approach. While accurately reflecting their parent contracts, the ECS and ECSSC also give contractors freedom to tailor more appropriately the needs of the subcontract by maintaining or varying the parent contract main and secondary options in the subcontract. They are probably closer to a contractor's needs than GC/Works equivalents.

Table 4.7 analyses some key issues regarding subletting under the two stables of contracts.

Table 4.7.

Key issue	GC/Works/1: Parts 1–3 and GC/Works/3	GC/Works/2 and GC/Works/4	NEC	NEC/ECSC
Extent of subcontracting	Partial	Partial	0–100%	0–100%
Consent to subcontracting required	Yes	Yes	No	No
Design of work by subcontractor	Yes – extent varies	No	0–100%	0–100%
Project Manager's requirement for subcontractor's details	Optional	No	Name only	No
Project Manager's requirement for subcontract details	Optional – may be prescribed in the Abstract of Particulars	No	Only if ECS or PSC not used or Project Manager agrees no submission required	No
Nomination provisions	Yes	No	No – restrictions included in Works Information	No
Management contract	Separate forms GC/Works/1: Parts 6 or 7 (construction management trade contracts)	Not applicable	Option F used in conjunction with ECS or ECSSC	Not applicable

Dispute resolution

Current GC/Works contracts have removed the words 'final and conclusive' which applied to certain decisions in earlier forms. These decisions could not be appealed to arbitration. Now, in specified circumstances, an aggrieved contractor is given the right to appeal against either Project Manager or quantity surveyor decisions. Outside these provisions disputes are resolved by means of adjudication, arbitration or litigation. Disputes under NEC contracts are referred to adjudication for resolution. On rare occasions when either party does not accept the result of the adjudication the dispute thereafter is referred to the chosen tribunal, i.e. arbitration or litigation.

Conclusions

No form provides a complete panacea for all the construction industry's needs. It is important, however, that for any particular project the chosen contract with its conditions fits the project, rather than the project forced to fit the contract.

Latham required a basic set of principles on which modern contracts could be based and a complete family of interlocking documents. GC/Works contracts, despite their multiplicity, each target a particular sector of the construction market and miss many other procurement options. NEC contracts, with their greater flexibility, have in contrast identifiable benefits over GC/Works contracts, cover a wider spectrum of the market and fulfil Latham's aspirations. Latham's 1994 recommendation that a target of one-third of government-funded projects started over the next four years should use the NEC may not have been achieved, but government-sponsored bodies are increasingly turning their attention to its benefits.

Twenty-first century contracts need to be robust enough to accommodate traditional and new initiatives, including private finance initiative, best value, prime contracting and supply chain management, etc. The NEC is well placed to meet the challenge. GC/Works contracts with their verbosity, straight-jacketed and somewhat staid approach are not similarly prepared.

NEC
compared
and
contrasted
with IChemE

Ernie Bayton

5

Introduction

Working groups established by the Institution of Chemical Engineers (IChemE) and the Institution of Civil Engineers (ICE) began with very different objectives when producing their respective documents for project procurement.

The IChemE addressed a specific type of work in the construction market, namely process plant. Launched in 1968 the *Model Form of Conditions of Contract for Process Plants Suitable for Lump Sum Contracts in the United Kingdom* (the 'Red Book') was the first of a number of model forms, each catering for different procurement criteria. Initially published for use in the UK using English law, the forms, with suitable amendment, can be used under other jurisdictions.

Following three key objectives New Engineering Contract (NEC) drafters adopted a multidisciplinary approach including civil engineering, electrical, mechanical and building work in a global market. They radically redefined contract strategies for procuring work. Substantial differences exist between IChemE and NEC. Practitioners will have their own preferences, but distinct advantages are available under the NEC.

Overview of IChemE and NEC

Current IChemE documents, produced over the past 10 years, reflect an evolving process. While each model form is published to address a different procurement option, other unrelated differences including terminology, definitions, procedures, obligations, etc. appear. As a group of documents, therefore, a disappointing lack of harmony exists between them. The published documents would benefit from an across-the-board revision removing unnecessary differences.

NEC documents are an interlocking family of contracts embodying the same three founding principles of flexibility, clarity and simplicity with the stimulus for good management. A common approach is also taken with regard to structure, options and content. The range of documents is more extensive than IChemE and where direct comparisons exist (e.g. construction main and short contracts), NEC contracts offer a wider range of options. The NEC range also includes a partnering option, professional services contract, short subcontract and flow charts accompanying all guidance notes, none of which are available within the IChemE stable.

Application

Although designed for use in the process plant industry, IChemE contracts are also used in other process-related industries. Their application and use however requires discretion. Process plants often include building and civil engineering work. Where process plant forms the main element of the work, IChemE conditions may be suitable. If building or civils work forms a substantial element either:

• special conditions should address that work
• procurement should be under a separate contract, or
• different conditions used for the entire project.

Parties dissatisfied with other standard forms have used the IChemE as an alternative, not always remembering the purpose for which they were designed. To those same parties the NEC provides a further, more flexible, alternative.

Allocating financial risk

All contracts carry financial risks. However, risks are allocated differently according to the procurement option chosen. Factors determining the apportionment of risk include any client desire for flexibility in their requirements post-award, the desired level of price certainty and stimulating efficiency within the organization providing the work or service. Figure 5.1 relates these three factors with differing styles of contract.

IChemE model forms are each stand-alone documents addressing different styles of contract. Red Book conditions are for lump sum contracts, Green Book conditions are reimbursable contracts covering target cost and cost plus options and Orange, Yellow and Brown Book conditions may be either. The Red Book may be used if part of the work is fully definable at tender stage and the balance sufficiently definable for reimbursable purposes. Fully defined work is priced as a lump sum and the balance priced as reimbursable. Reimbursable elements, once defined, become lump sums eventually creating an entire lump sum contract. Green Book conditions envisage a flexible approach towards reimbursable contracts.

NEC contracts differ considerably, using common core clauses supplemented by main options dealing solely with the method of payment. Varying the main option alters the allocation of financial risk between the parties – see Table 5.1. Secondary options are available for use in conjunction with the core and chosen main option clauses.

NEC contracts and Figure 5.1 factors combine enabling the correct choice of main option for the contract – see Table 5.2.

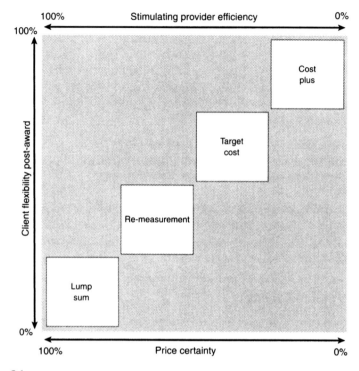

Figure 5.1.

Table 5.1.

Main financial risk bearer	NEC contract – main option		
	PSC	ECC	ECS
Employer	E	E & F	
Consultant/contractor	A & G	A & B	E
Subconsultant/subcontractor	A & G		A & B
Shared between the parties	C	C & D	C & D

Design

Red and Green Book agreements refer to the Contractor's obligation to 'complete the design … [of the Works]'. The extent of work required to complete the design will vary from contract to contract. More purchaser design pre-tender correspondingly reduces the amount of design input to be done at tender submission or post-award stages by the Contractor. The wording of the agreement is significant and indicates that the Contractor does not have total responsibility for all the design for the works. The general conditions, apart from a reference at clause 3 in both model forms enabling

Table 5.2.

Contract style (Figure 5.1)	NEC contract – main option		
	PSC	ECC	ECS
Lump sum	A	A	A
Re-measurement	G[1]	B[2]	B[2]
Target cost	C	C & D	C & D
Cost plus	E	E & F	E

[1] Consultant's risk is the ability to perform Employer instructed tasks at agreed rates.
[2] If bill of quantities are definitive, option B effectively becomes a lump sum priced contract, not subject to re-measurement.

modification of the design at the Contractor's initiative, contain no express provisions obliging the Contractor to accept retrospective responsibility for any design work undertaken by the purchaser. These principles need to be set against the contractor's obligation that the plant (not the works) are to be, 'in every respect suitable/fit for the purpose/s for which it is intended' (Red/Green Books, respectively). Drafters of IChemE contracts must necessarily exercise great care in defining within the contract itself the purpose for which the plant is intended and what it requires to achieve.

Orange, Yellow and Brown Book agreements make no specific reference to design by the Contractor or subcontractor. Use of the Orange Book is probably inappropriate where design is to form a part of the contractor's work. Design by subcontractors must be specified in schedules attached to the subcontracts.

All NEC construction contracts facilitate 0–100% design responsibility by the Contractor or subcontractor, the Works Information setting the limits of responsibility. They differ from IChemE contracts in two important respects. First, the contractor's obligation is to provide the works in accordance with the Works Information. That obligation may take the Contractor closer to 'fitness for purpose' than to 'reasonable skill and care'; hence the option is provided for incorporating secondary option M into the contract (limitation of the contractor's liability for design to reasonable skill and care). The second key area is the availability of a NEC professional services contract (NEC/PSC). The NEC/PSC fully interlocks with the NEC matrix of documents and is capable of use by purchasers, contractors and subcontractors alike with their respective consultants.

Administration

Administrative procedures vary between the two groups of contracts in a number of respects. Table 5.3 identifies the key areas where particular differences exist between comparable contracts in the two groups.

Table 5.3.

Subject	IChemE contracts			NEC contracts	
	Red	Green	Orange	ECC	ECSC
Management					
Employer's agent	Project Manager	Project Manager	Purchaser's Representative	Project Manager and Supervisor[1]	Employer acts for themselves
Delegation by notice	Yes – to Project Manager representative	Yes – to project manager representative	No	To delegate	To delegate
Limits on delegation	Yes	Yes	Yes	No	No
Contractor's agent	Contract manager	Contract manager	Contractor's Representative	Key people[2]	No
Limits on powers	No	No	No	No	No
Appointed deputy	Optional	No	No	Key people[2]	No
Appointed site manager	Mandatory	Mandatory	No	Key people[2]	No
Deputy site manager	If site manager absent	If site manager absent	No	Key people[2]	No
Documentation					
Contract documents defined	Yes	Yes	Yes	Stated in agreement[3]	No[4]
Priority of documents specified	Yes	Yes	Yes	No[5]	No

Procedures

Procedures				
Reply period for communications stated	Yes[6]	Yes[6]	No – only under certain provisions	No – only under certain provisions
Notices to be communicated separately from any other communications	No	Yes	No	No
Early warning procedure	Yes	Yes	No	No
Progress meetings	No	No	No	Yes
Acceleration provisions	No	Yes	No	No
Detailed testing and inspection procedures	No	No[7]	Yes	Yes
Specified period within which a defect is to be remedied	Yes	Yes	No	No

[1] The Project Manager and Supervisor undertake different roles.

[2] Key people are chosen and identified by the Contractor in the Contract Data.

[3] By stating the documents forming the contract in the agreement the parties avoid having to amend contract conditions to cater for documents not named.

[4] The published ECSC documents establish the contract.

[5] Project Manager is free to prioritize according to differing circumstances.

[6] A common period for reply is stated for all communications, varied only by specific clauses adopting discrete timetables.

[7] Testing and inspection procedures are set out in Works Information rather than the conditions.

Programmes

Modern contracts are executed within strict time constraints. An essential tool for achieving the timely delivery of projects is the programme. Good, reliable and accurate programming procedures capable of adequately monitoring and controlling events are necessary. Agreed programmes should enjoy all-party support.

IChemE contracts contain rules for programme submission, approval, compliance, revision and relevance where delays occur. Their provisions when compared with NEC contracts are however remarkably brief and less robust. NEC contracts are programme driven to a greater extent and key advantages gained from their provisions are:

- better definition of programme content
- detailed supportive documentation specified
- revision frequency specified
- revision details better defined
- acceptance procedures better defined
- financial penalties upon contractors failing to submit a first conforming programme
- a specified role in identifying some compensation events
- a requirement to provide programmes as a part of quotations for compensation events
- planned completion is used to assess delay entitlements.

Payment

Lump sum contracts

Contracts carried out under Red Book conditions are lump sum contracts, the contract price being subject to adjustment by way of variations, allowable delay costs, etc. The pricing document by which the contract price is established is not specified in the general conditions. Some degree of analysis is required however to facilitate interim payment entitlements (detailed in Schedule 8) and payment for variations (rates and charges detailed in Schedule 11). Guidance is provided regarding the formation of these two schedules although it is left to the parties to establish their own arrangements for each contract.

NEC priced construction contracts (ECC and ECS) provide two, well-defined, pricing mechanisms, activity schedules (option A) and bill of quantities (option B), responsibility for their production and the associated risks lying with the Contractor and purchaser, respectively. The respective merits of each require to be assessed in the light of the project prior to tender invitation. Advantages of activity schedules include:

- tenderers control its detailed content
- purchasers may specify the activity schedule outline (enabling tender comparisons)
- activities may be grouped or listed singly
- activities may relate to actions either at a specific point or over a period of time
- activities can be priced according to the resources required for their procurement
- activities can account for natural breaks in the construction process
- activities can account for different working methods for similar types of work
- activities can account for similar work undertaken at different times
- activity prices are lump sums not unit rates
- activities are related to the programme
- payment is related to the completion of activities (either as a group, or singly where not in a group)
- cash flow forecasting is easier than with a bill of quantities.

Changes in NEC priced contracts are dealt with as compensation events. The Contractor provides a quotation for each compensation event based on their forecast of their costs of executing changed work. An exception applies to option B contracts, where rates and lump sums in the bills of quantities may be used as an alternative, subject to the agreement of the Project Manager and Contractor.

Reimbursable contracts

Ostensibly IChemE Green Book (reimbursable) contracts have greater flexibility than NEC construction contracts since they are not limited to re-measurement, target, cost or management options. Additionally they have the facility to include lump sum and reimbursable elements, although problems can arise when separating costs for each element. In practice the options are less wide than with NEC construction contracts and tend to narrow down to:

- cost plus a percentage fee
- cost plus a fixed fee
- target
- guaranteed maximum price.

NEC options use two mutually exclusive mechanisms for establishing the target contract prices: activity schedules (Option C) and bill of quantities (Option D). Additionally there are cost reimbursable (Option E) and management contract (Option F) options.

Payment for work under Green Book contracts is made against schedules of cost elements and rates and charges. Guide notes set out typical items for inclusion. NEC contracts similarly have schedules of cost components, but their content is better defined and is incorporated into the conditions. Actual cost and disallowed cost are defined terms.

Changes are dealt with as compensation events, the purpose of the agreed monetary evaluation differing according to the main option used.

By comparison with IChemE contracts, NEC reimbursable contracts (options C–F) are more prescriptive regarding a contractor's obligation to prepare, at agreed intervals, forecasts of the total actual cost for the works.

Compensation events

Compensation events under NEC contracts bring together both time and monetary entitlements in a single clause, thus avoiding separate notices, procedures, entitlements, etc. The IChemE conditions adopt a different approach. Key differences are analysed in Table 5.4.

Subcontracting

Twenty-first century contracting involves subcontracting substantial parts of the work by contractors. Subcontracting under IChemE conditions is subject to various limitations. The detailed provisions relating to subcontracting differ depending upon the Model Form used, although the reason for the differences does not always appear to be related to the different purpose each model form serves. NEC subcontracting procedures are more uniform and differ from IChemE Forms in the following ways:

- no nomination provisions exist – where purchasers wish to control subcontractor selection alternative options are available such as direct contracts, naming or select lists, in the Works Information, etc.
- Project Manager acceptance of the proposed subcontractor is required
- Project Manager acceptance of the proposed subcontract conditions is required, unless an NEC document is to be used, or the Project Manager relaxes the requirement
- principles of mutual trust and co-operation must be stepped down into subcontracts
- NEC subcontract conditions more accurately mirror parent contract conditions than with IChemE subcontracts
- payment options can vary from the parent contract and between subcontracts

Table 5.4.

Key issue	IChemE (Red & Green Books)	NEC/ECC
List of events	Fewer reasons but they include 'force majeure' – a term defined in the conditions	18 reasons are given in the core clauses; others arise under main and secondary options; additional reasons may be introduced by employers at tender stage
Notification	By Contractor except force majeure which may be notified by either party; some clauses require a notice as a condition precedent to any entitlement	By Project Manager in certain circumstances, otherwise by Contractor; notices to be given timeously and the defects date acts as a time bar
Linkage of time and monetary entitlements	Varies – depending upon the reason for delay, monetary entitlements may: • not accrue at all • be reimbursed under the delay clause • be reimbursed under another clause • attract value not cost	Both time and money are considered for their effect upon actual time and actual cost
Provision of quotations	Procedures are less well defined than for NEC	Quotations to comprise time, money, supporting details and, where appropriate, programmes; the purpose of the quotation differs according to the main option used
Timetable for quotations	Less well defined than NEC	Timetable applies to Contractor and Project Manager for submission and acceptance, respectively *(continued)*

Table 5.4. (*continued*)

Key issue	IChemE (Red & Green Books)	NEC/ECC
Monetary evaluation	Varies and may attract value or cost	Changes to prices are based upon the actual cost as defined by the main option used; contracts using bill of quantities may vary this rule
Role of programme	Variation issued covering the extension to the Approved Programme and Time for Completion (Red Book); Green Book conditions are less specific but relate to the programme of work	Delays are assessed against planned completion on the accepted programme
Events too uncertain to be forecast reasonably for quotation purposes	No specific provisions	Project Manager to specify assumptions for basis of quotation; if proved wrong the quotation is corrected
Project Manager's ability to assess event him/herself	No specific provisions	Project Manager may assess entitlements in prescribed circumstances

Table 5.5.

Method/form	Red book	Green book	Orange book	Yellow book	Brown book	Comments
Certain project manager representative's actions appealed to Project Manager	X	X				This procedure may be invalid if HGCRA provisions apply to the contract
Negotiation	X			X	X	Always available whether mentioned in the contract or not
Dissatisfaction		X	X		X	This procedure may be invalid if HGCRA provisions apply to the contract
Mediation	X			X	X	Optional procedure; not available under NEC
Expert	X	X		X	X	Conditions define certain matters to be referred to an Expert, others are referred by mutual agreement; normally used where a speedy resolution is required; decision cannot be appealed to arbitration
Adjudication			X	X	X	A draft amendment is available for conditions published pre-HGCRA; not all process plant work falls within the terms of the Act, the definition of 'Construction Operations' excludes some types of work carried out under IChemE Forms.
Arbitration	X	X	X	X	X	Only some conditions specify the arbitration rules to be adopted
Third Party			X			Gives considerable freedom to contracting parties but decision is final, binding and not subject to review
Litigation	X	X	X	X	X	

- secondary options can vary from the parent contract and between subcontracts
- subcontract conditions are suitable for all types of subcontract work
- the short subcontract is available and is compatible with both the NEC and the short contract
- the PSC is available for consultancy subcontracts
- the PSC addresses subconsulting.

Contractors will find distinct benefits using NEC subcontracting arrangements *vis-à-vis* IChemE procedures.

Dispute resolution procedures

All contracts require adequate procedures to deal with disputes should they arise. The IChemE forms between them boast nine options, used in different combinations, for addressing disputes. The comparative options by which disputes may be resolved under each model form are set out in Table 5.5.

IChemE's multiple approach is disadvantaged by variances between the model forms, some of which are mandatory and others optional, some common and others not.

Key differences under the NEC are that:

- fewer methods of dispute resolution exist
- a common approach is used across all NEC contracts
- all disputes are referred to adjudication
- a strict timetable applies
- adjudicators are named in contracts
- adjudicator's decisions if not accepted are referred to a pre-determined tribunal, arbitration or litigation
- adjudicator's fees are always shared equally between the parties
- Housing Grants, Construction and Regeneration Act 1996 requirements are covered by a secondary option.

Conclusions

It is essential that users of IChemE forms keep in mind the purpose for which they were designed. As a group, some disharmony exists across the range of documents that can be frustrating for practitioners. Since their introduction, the forms have increased in popularity and served the process plant industry well. Without some revision however (revisions are due in 2002), recognition of the changing initiatives by which work is procured (private finance initiative, prime contracting, etc.) and a review of the available documentation, the IChemE forms may well lose ground in today's market. The forms as presently drafted do not meet Latham's requirements for a basic set of principles on which modern contracts could be based and a complete family of interlocking documents.

NEC contracts in contrast readily accommodate the special demands of the process plant industry, are suitable where a mix of disciplines exist, meet Latham aspirations and are robust enough to meet twenty-first century challenges.

NEC
compared
and
contrasted
with MF/1

Nigel Shaw

6

Introduction

This chapter compares the New Engineering Contract (NEC) with a form of contract that is frequently used for the supply and erection of all forms of electrical, electronic and mechanical plant, namely form MF/1. This is a contract published jointly by the Institution of Mechanical Engineers (IMechE) and the Institution of Electrical Engineers (IEE). It is endorsed by the Association of Consulting Engineers and often recommended by electrical and mechanical engineering consultants.

MF/1 has evolved over many years. It is the current version of a form first developed as early as 1903, when what must have been a very progressive committee of the IEE, decided that the new technology needed its own form of contract.

The contracts being compared are the NEC Engineering and Construction Contract, 2nd Edition 1995 (NEC) and the model form MF/1 General Conditions of Contract, 2000 Edition (MF/1).

Contract philosophy

The objective of the NEC family of contracts is to stimulate good management, to be useable in a wide variety of situations and to be clear and as simple as possible in words and structure.

The objective behind MF/1 is not clearly set out, perhaps because the authors think this goes without saying, but appears to be no more than to provide a standardized, fair and equitable contract to govern the relationships between the parties.

The application of these 'philosophies' results in significant differences in both content and style – see Table 6.1.

The MF family of contracts

The Joint IMechE/IEE MF series of model forms consist of:

- MF/1 (Revision 4) 2000 Edition. Home or overseas contracts for the supply of electrical, electronic or mechanical plant – with erection (it includes a form of subcontract).
- MF/2 (Revision 1) 1999 Edition. Home or overseas contracts for the supply of electrical, electronic or mechanical plant (it includes a form of subcontract and a form of supervision contract).
- MF/3 1993 Edition. Home contracts for the supply of electrical and mechanical goods.
- MF/4 (in preparation). Terms and conditions for the engagement of engineering consultants.

Table 6.1. *Compared with the NEC the MF/1 is a traditional form*

NEC	MF/1
Created 10 years ago to provide a modern tool to manage all types of engineering and construction	Created in 1903, with last major revision in 1988. An evolved contract adapted to, but not always up to date with, generally accepted best practice
The Project Manager appointed under the contract manages the work for the Employer as agent. The Project Manager has no independent role to arbitrate or judge between the parties	The Engineer manages the work and while appointed by the Employer, has a duty to act fairly between the parties, when, for example, valuing variations
Drafted to achieve clarity and in straightforward language. It uses the present tense throughout	Drafted in conventional legal language with the use of 'shall' to impose obligations

Assembling a contract

At the project definition stage a key procurement decision is the pricing strategy for the contracts needed to construct the project. NEC users have a wide choice from lump sum prices to a management contract via target, reimbursable and bill of quantities options. All these are included in the six main pricing options included in the NEC contract.

MF/1 users are restricted to lump sums or, by means of a set of standard amendments, bills of quantities.

The NEC guidance notes discourage amendments. The contract has an option allowing *additional* clauses, but it is recommended that only those absolutely necessary because of the nature of a contract are introduced, and amending existing clauses is strongly discouraged. By contrast MF/1 anticipates that additional special conditions will be used and provides some model clauses to, for example, allow payment by measurement.

A comparison of how the structure of an NEC contract may compare with its MF/1 equivalent is shown in Table 6.2.

Managing a contract

While the differences in the documentation may affect the way in which a contract is prepared (and the cost of drafting) the real test of a contract is in the way it delivers a project.

Table 6.2. *Comparison of the structure of an NEC contract with its MF/1 equivalent*

NEC	MF/1
Instructions to tenderers	Instructions to tenderers
Form of tender	Form of tender
NEC core conditions of contract Contract Data Part One – data provided by the Employer. All the Employer data needed to manage the contract from the main pricing option to the dispute tribunal chosen and any additional contract clauses	MF/1 general conditions of contract General conditions appendix. Values required by contract clauses, e.g. prime cost percentage and arbitrator appointing body Special conditions. Balance of Employer provided data needed to manage the contract, together with a list of additional special conditions Additional special conditions
Contract Data Part Two – data provided by the Contractor. All the contractor data needed to manage the contract	Appendix or schedule to form of tender, completed by the tendering contractor
Activity schedule Bill of quantities Site Information Works Information	Segregation of contract price Rates schedule Specification. (Note that some matters that would be dealt with as additional special conditions in MF/1 would be Works Information in the NEC)
Both contracts may require other schedules and forms, for example bonds, forms of agreement	

Someone deciding the choice of a contract could refer to line-by-line commentaries on such a form, but even for those possessed with an adequate library, the decision will hinge on a consideration of the prime differences. Table 6.3 identifies what many would consider these differences to be.

What really matters?

In *Constructing the Team*, Sir Michael Latham strongly advocates the use of an NEC contract because he saw it as providing the best framework to improve the way in which construction is managed. However, some non-construction users of other forms of contract including MF/1 saw no need to change, on the basis that they did not have problems with existing forms. A few – from all 'sides' of a project – were bold enough to say 'We do not have a confrontational culture in our industry, and the contract stays in the drawer'.

Table 6.3. *Deciding between the NEC and MF/1*

	NEC	MF/1
The contracting culture	The language, layout and management systems are unquestionably 'modern' in approach and commitment to best practice. It will therefore particularly appeal to clients who are actively pursuing best value for money and continuous improvement	Based on traditional relationships, defensive risk management and seeks conservative, incremental improvement. It will appeal to clients who have not had problems in the past and are comfortable in using a document familiar to the contracting parties
Working together	Makes working together 'in a spirit of mutual trust and co-operation' an obligation. While it is difficult to force people to co-operate if they don't want to, this 'up-front' declaration in the first clause of the contract does seem to be effective in encouraging co-operative working	MF/1 does not have any equivalent clause and if users want to emphasize co-operation, they will need to amend the contract or write a partnering process into the contract
Managing the contract	(1) The Project Manager is the client's agent, and the client has total flexibility in how the project management is organized	(1) The Engineer acts independently under a number of clauses. This can limit flexibility and client control, and real contractor benefits are often illusory
	(2) The contract procedures are designed to encourage looking ahead to solve problems and reduce risks, on the basis that it is only possible to plan to manage what happens in the future, not what has already happened	(2) Contract procedures are based on traditional liability driven incentives to perform. There is no procedural obligation to jointly solve problems

(continued)

Table 6.3. (*continued*)

	NEC	MF/1
Certainty of outcome – no nasty surprises	(1) Early warning process requires either party to notify the other promptly if they become aware of something that will increase cost, delay the programme, or impair performance. This is an extremely powerful tool to ensure both parties have, at every stage, the best possible understanding of the likely outcome of the project	(1) There is no equivalent here and without such an obligation there is no need for either party to help the other. With goodwill this need not necessarily be a problem, but there is no incentive for the parties to exhibit goodwill. Several drafting bodies have 'borrowed' the NEC idea and included early warning. The omission of such a clause in MF/1 is not helpful
	(2) Variations (which are not limited by value) and claims have a common procedure in the NEC and are known as 'compensation events' (ce's). A logical process of notification, assessment, acceptance, and implementation is included. Compensation events are intended to be priced as soon as possible after they occur, in 'real time' so that for complex events both parties can participate in managing costs that have yet to be incurred. Once a 'ce' is accepted, the price is not normally varied, and hence both parties are aware of the final price as early as is possible	(2) Variations may be instructed provided the contract price does not change by more than 15%. Unless the contract has a relevant schedule of prices, the Engineer has to decide on a 'reasonable' price. If work is instructed before a price is determined, the Contractor must keep records. There is no requirement for the Engineer to value variations promptly and valuation may be delayed until the work is complete

(3) A list of acceptable compensation events lets both parties see where risks lie. 'Claims' that are not on the list of compensation events will not be valid. Ce's must be notified promptly and go through the same real time cost/method/ product assessment process as for variations. Thus both parties have the earliest understanding of final contract costs and can influence resolution of the problems created by the event leading to the 'claim'

(3) Claims are required to be notified to the Employer within 30 days of circumstances arising which may lead to the claim. The Contractor is then required 'as soon as reasonably practicable' after the notification, and not later than the end of the last defects liability period, to give full details of the claim. There is no method for assessing the claim provided and in practice details of the claim may often not be known by the Employer until the end of the contract

Payment

The NEC payment process assumes a regular assessment of the payments due to the Contractor, but the outcome from this varies between the main pricing options. Thus an Employer can choose between paying lump sums for completion of defined activities through to regular cost reimbursement. The target options provide regular cost reimbursement of cost incurred with an end of contract exchange of the resulting pain/gain share

MF/1 uses special conditions to allow interim and progress payment, or by measurement. It is presumed that employers will often need to amend the model text to meet their particular requirements. Fewer options are provided and these are less integrated into the contract than for the NEC

Force majeure

The NEC does not use or define 'force majeure' and except as provided by the compensation event procedure, risks rest with the party affected

Force majeure is defined and parties may be excused of performance

(continued)

Table 6.3. (*continued*)

	NEC	MF/1
Settling disputes	The NEC processes are specifically designed to prevent disputes before they arise, but if this fails an adjudication procedure is included [an amendment is used in the UK to provide Housing Grants, Construction and Regeneration Act 1996 (HGCR) compliance]. Options chosen by the Employer pre-contract allow 'appeal' to a tribunal. The tribunal may be any pre-selected form of dispute resolution (including arbitration). Many users choose to go straight to the courts, and select 'litigation' as the tribunal	If the Contractor disputes instructions, etc. of the Engineer, MF/1 provides for the Engineer to promptly respond with reasons. The response is binding unless referred to arbitration. Arbitration is provided for all disputes between Employer and Contractor. There is surprisingly no provision for adjudication except in the UK if the HGCR applies, when a special condition is available
General conditions vs the specification	Designed as a generic document and does not include industry or project specific clauses, which are provided as needed by the Employer in the Works Information (i.e. the specification)	MF/1 is for use with electrical, electronic, and mechanical plant. As such it contains a number of clauses that are technology specific

Anyone wanting to use a conventional (MF/1) specification as part of an NEC works information section or vice versa would need, for example, to review, the following matters:

| Testing | The NEC clauses provide for testing, but project specific requirements must be written in the Works Information, together with full technical details of the tests. The NEC core does not distinguish performance tests as such, as they are considered part of the generic 'tests', but a secondary option provides damages for low option performance | MF/1 has two key clauses dealing with tests: 'Tests on completion' provides for basic testing while 'Performance tests' gives the arrangements for deciding if performance criteria have been reached. While more detailed than the NEC, MF/1 still requires full technical details of the tests to be written in the specification |
| Software | The NEC, by intent, does not treat software as being distinct from other intellectual property required to allow a product to be used. If technology specific provisions need to be made these are 'Works Information' | MF/1 has additional special conditions for contracts with an incidental supply of software or hardware. Many of the 16 clauses are 'technical specification', with the remainder dealing with special title, testing, confidentiality, etc. |

Knowledgeable clients are not always so relaxed. Client procurement objectives, whether for construction or engineering, must be to get the best value for money. The way in which the conditions allow, or hinder, the best system of management will affect success. *Status quo* is never an option if there is a better way, unless the project is so small that it is not worth the effort of diverting time from more profitable tasks to effect changes.

Thus a client, particularly a 'serial' client, and even more particularly one already using the NEC family for construction work, would gain worthwhile advantages in using that system for all engineering procurement. While both client and contractor may have gained some comfort that they knew how to get round problems that might arise on an MF/1 contract, both would soon learn that the NEC offers real (cost, delivery and performance) advantages.

Conclusions

MF/1 is well established and has many users but has not yet embraced contemporary procurement 'best practice'. It could be seen to offer few benefits to clients or contractors except possibly for familiarity and the inclusion of terms specifically relating to plant testing. It will obviously appeal to those who support a liability driven contract but does not facilitate a purchaser needing to actively manage and improve the performance of a supply chain.

Having said that, the traditional approach of MF/1 may still work satisfactorily if the suppliers and purchasers concerned have established processes that compensate for the absence of the most modern management methods.

The NEC is a very different document, with many advantages, not the least of which is its clear encouragement of good management.

The pro-active problem solving approach of the NEC means that the project team should most certainly not leave it in the drawer, but should use its processes to prevent the many problems that may have otherwise arisen, and to minimize the effect of those that cannot be prevented.

However, the approach of the NEC is sufficiently different from the traditional to the extent that its full benefits will not be fully exploited, unless the project team leaders have a degree of training, and an adequate understanding of its philosophy and application.

In conclusion, the NEC is still relatively new, but is proven in use and will be the choice for projects that need a process that rises above the average.

PSC compared and contrasted with RIBA SFA/99

Frances Forward

7

Introduction

The objectives of this chapter are first to compare the basis of the New Engineering Contract (NEC) Professional Services Contract (PSC) and the Royal Institute of British Architects (RIBA) standard appointment documents and secondly to analyse those aspects which display significant differences in their rationale and operation. While the PSC is designed to be used for a multiplicity of consultant appointments, the RIBA standard appointment documents are expressly intended for appointing architects in their 'conventional' roles. This comparison is therefore specifically in relation to employing architects in all their potential capacities as designers, 'design team leaders' and building contract administrators, but does not look at the appointment implications of other potential roles.

Nature of the project

Standard forms of professional services contract have been based conventionally on the principle that the size of project dictates the standard form version used. The underlying rationale for this approach appears to rely on the notion that complexity increases with size and that smaller projects require simpler contracts (see Table 7.1).

Table 7.1.

Size of project/degree of complexity	Small	Medium	Large
Standard Forms			
RIBA SFA/99	–	–	✓
RIBA CE/99	–	✓	–
RIBA SW/99	✓	–	–
PSC 98	✓	✓	✓

Abbreviations: RIBA SFA/99: RIBA Standard Form of Agreement for the Appointment of an Architect 1999
RIBA CE/99: RIBA Conditions of Engagement for the Appointment of an Architect For use with a Letter of Appointment 1999
RIBA SW/99: RIBA Small Works Conditions of Appointment & Schedule of Services 1999
PSC 98: NEC Professional Services Contract: 2nd Edition 1998.

Certainly, the 1999 suite of RIBA forms of appointment follows such logic: SFA/99 being intended for larger, more complex, projects, incorporating articles of agreement; CE/99 being intended for 'medium' size/ complexity projects, where a letter of appointment is preferred and SW/99 being intended for relatively straightforward projects under £150,000 in construction cost, using a letter of appointment. Conversely, the PSC is founded on the principle that the contract conditions should offer the same flexibility, irrespective of size or complexity of projects, particularly as the two attributes are not necessarily connected.

Key issues

Both the PSC and the RIBA standard appointment documents share the objective of setting out the parties' respective rights and obligations. Of paramount importance in this regard, to architects and employers alike, is first clarity in relation to the services to be provided by the Architect, and secondly certainty of fees to be paid by the Employer for the provision of such services.

Role of the RIBA Plan of Work

The RIBA Plan of Work (RIBA Publications 2000) is not only enshrined in the methodology of most UK architects and their clients for managing the content and progress of design and construction, it also forms the basis of breaking down the Architect's services for the purposes of correlation with fees. Furthermore, it provides a benchmark for other consultants involved in building design – whether directly appointed by the client, or appointed as subconsultants to the Architect – to measure their performance and fees. The absence of any express reference to the RIBA Plan of Work in the PSC might, therefore, be perceived as a failing in relation to UK project management; this would, however, be a fallacy, as there is implicit provision to include it in the Scope (core clause 11.2(5)) that forms an essential part of the PSC documentation. There is also provision to link the payment of fees to completion of Activities (main options A and C), which can be RIBA Work Stages, or parts thereof; i.e. the activity schedules envisaged in the RIBA Plan of Work can form the fee proposal/payment pricing document under the PSC.

Services

The RIBA standard appointment documents provide a standard list of services associated with each work stage in the RIBA Plan of Work. The PSC is predicated on the principle that there is a need to define a project specific range of services, in order to respond to the prototype nature of most construction projects (see Table 7.2).

Table 7.2.

Form of contract	RIBA SFA/99	RIBA CE/99	RIBA SW/99	PSC 98
Services definition				
RIBA Plan of Work services A/B – with/ without cost advice	✓ (Schedule 2)	✓ (Schedule 2)	✓ (Service Schedule)	✓ (Scope)
RIBA Other Services – Activities/Special activities	(✓) optional	(✓) optional	(✓) optional	(✓) (Scope)
Entirely bespoke services	–	–	–	✓ (Scope)

Fee basis

Table 7.3 summarizes the standard methods of fee calculation under the comparative forms.

Table 7.3.

Form of contract	RIBA SFA/99	RIBA CE/99	RIBA SW/99	PSC 98
Fee basis				
Variable percentage	✓ (cl.5.2)	✓ (cl.5.2)	✓ (cl.18/19)	–
Lump sum	✓ (cl.5.3)	✓ (cl.5.3)	✓ (cl.18)	Main option A
Time charge	✓ (cl.5.4)	✓ (cl.5.4)	✓ (cl.20)	Main option E
Target	–	–	–	Main option C
Term	–	–	–	Main option G

The common allocation of resource risk between the parties where architects carry out RIBA Work Stages A and B on a time charge basis, while the project is defined, and subsequently carry out RIBA Work Stages C–L on a lump sum fee basis, can be accommodated under the PSC by means of sequential use of the time-based payment provisions of main option E and the priced activity schedule payment provisions of main option A. Alternatively, resource risk can be allocated under the PSC on a reciprocal incentivized basis by means of the target provisions of main option C and the resulting 'pain or gain' share between Architect and Employer for fees above or below the target costs could be based on different percentages for RIBA Work Stages A and B and RIBA Work Stages C–L, to better reflect the parties' sequential ability to carry and manage such risk. In the context of a recent resurgence of earlier debates among employers, and to some extent architects, as to the appropriateness of calculating fees by applying percentages to projected construction costs, the emergence in the PSC of more flexible, creative, equitable and even incentivized fee options can only be welcomed. While there is no mandatory link between the use of the RIBA standard appointment documents and the recommended fee scale published by the RIBA, and while virtually all architects apparently significantly undercut the RIBA recommended percentage and time charge fees in an effort to remain competitive, it would seem more likely that accurate and efficient resource-based fee proposals will emerge from application of the PSC main options than from application of the RIBA standard appointment documents. Strategic and project-specific planning is perhaps more crucial to a PSC-based fee proposal than an RIBA-based one; however, the benefit of the RIBA Plan of Work is probably even greater.

Contract machinery

In addition to the different approaches to services definition and basis of fees, there are a number of other significant differences in the contract machinery to control important concepts.

Programme

The RIBA standard appointment documents refer to the Project Timetable, which, while it may exist as a precise programme, in practice is often more of a client 'wish list', the 'enforceability' of which may be perceived differently by the two parties. The PSC, in contrast, makes the programme an integral contract document, which is kept up to date and can be used as a project management tool, both for monitoring progress accurately and in relation to incentivization. Many architects may feel uncomfortable about performance of their professional services to a strict programme, with 'nowhere to hide' and possibly even at risk of delay damages, if secondary option X7 is incorporated. However, where a resource-based fee proposal forms the basis of the professional services contract and in the context of clients and contractors increasingly blaming project delays on late consultants' information, the PSC programme requirements can be seen as a valuable aid to realistic performance objectives, which in turn could enhance both profitability and reputation of both parties.

Change control

The power to order change is seen by most employers in the construction industry to be an essential provision in their professional services contracts, as well as their building contracts. Both the PSC and the RIBA standard appointment documents have procedures designed to manage change: the so-called 'Compensation Event' procedure in the PSC, which is mirrored in the NEC, and Services Variation/Additional Fees provisions in the RIBA standard appointment documents. There are, however, important procedural differences between the contracts. The RIBA standard appointment documents impose a reciprocal duty to advise of necessary variations and require the parties to 'agree' action; this may be regarded as contentious and subjective in some instances, particularly as no time limit is imposed for such agreement and no critical path is likely to have been identified in the Project Timetable. The only methodology envisaged to deal with the fee implications of such Services Variations is a time charge basis. The PSC, in contrast, makes early assessment of cost and time implications of all changes both mandatory and final, and crucially, it requires the fee and programme implications of compensation events to be related to the agreed contractual fee basis and the detailed contract programme. This will arguably not only achieve earlier certainty, but also fairer assessment relative to the parties' original expectations prior to such compensation events.

Design and Build procurement

The arrangements for 'Design and Build' procurement under the RIBA forms of appointment are based first on the principle of 'Amendments' (DB1/99 and DB2/99) to the traditional arrangements and secondly on a somewhat subtle interpretation of the distinction between 'Consultant Switch' and 'Novation'. The PSC makes the transition from 'traditional' to 'design and build' procurement much more effortlessly, by virtue of the flexible content of the Scope. This difference is best understood in the context of the RIBA forms of appointment attempting to respond to the not entirely compatible JCT Standard Form of Contract With Contractor's Design 1998, with its 'Employer's Requirements' and 'Contractor's Proposals', while the PSC can rely on compatibility with the NEC building contract, as a result of its inherently back-to-back drafting.

Quality

The RIBA forms of appointment make no express reference to quality requirements, but rather rely on an established understanding of both the statutory Architects Registration Board (ARB) Code of Conduct for all registered architects and the RIBA Code of Conduct for RIBA members, as well as the common law requirement to act with 'reasonable skill and care', as measured in relation to other competent members of the profession. The PSC does not alter any of those obligations; it does, however, require consultants to identify a quality management system for providing the services. This is arguably a benefit to client and architect alike, in terms of applying best practice and monitoring performance.

Dispute avoidance and dispute resolution

The RIBA forms of appointment comply with the statutory adjudication requirements of the Housing Grants, Construction and Regeneration Act 1996 (HGCR Act) and now offer a choice of tribunal, as between arbitration and litigation. They also makes provision for alternative dispute resolution in the form of negotiation or conciliation, but currently contain no provisions for partnering arrangements. The PSC, like the NEC, is founded on the fundamentally non-adversarial principle of the parties acting 'in a spirit of mutual trust and co-operation' and on the requirement for the parties to give reciprocal 'early warning' (core clause 15) of matters which could prejudice time, cost or quality parameters, thus enshrining the first ingredients of partnering. The PSC also provides for 'non-statutory' adjudication for domestic projects not under the HGCR Act and for international projects, with separate statutory provisions for those UK projects which do fall under the HGCR Act (core clause section 9 and secondary option Y(UK)2).

Application

The 1999 suite of RIBA forms of appointment is designed to be used for the appointment of architects, employed by clients or contractors, or by clients and subsequently contractors. Allowance is made for architects to employ subconsultants, by means of the SC/99 form, which refers to the 'normal services' of the subconsultant's profession. In contrast, the structure of the PSC, with the crucial role of the Scope in defining the services to be performed, allows back-to-back 'nesting' of a potentially infinite number of PSCs, without the need for reconciliation of a variety of generic services conventions, evolved independently through the professional institutions. The emphasis with the PSC and the role of its Scope is always focused on specific needs of individual projects.

The 1999 suite of RIBA forms of appointment is designed to be used for 'domestic' projects, i.e. essentially in England and Wales only, albeit with Royal Incorporation of Architects in Scotland replacement pages for use in Scotland. The PSC, in contrast, is designed to be operable internationally and there are useful provisions for international employers, e.g. the option of multiple currencies (secondary option X3) and the ability to choose the law of the contract (Contract Data Part One).

There is a tendency to correlate the use of the suite of RIBA forms of appointment with the JCT family of building contracts and certainly the back-to-back drafting principles of the PSC achieve an unprecedented compatibility when used with the NEC building contract and sub-contracts. If, however, project circumstances dictate, there is no inherent difficulty in mixing forms from different drafting bodies and eras, as between professional services and building contracts, provided their respective principles are well understood.

Overview

A summary of how the PSC can offer better overall project management control is shown in Table 7.4.

Conclusion

There is clearly no such thing as a perfect method for appointment of architects and therefore no perfect standard form of professional services contract for a particular project; there are also professional ethics and consumer politics issues at stake in this sphere. Ultimately, the most important aspect of the contractual relationship between client and architect is likely to be common expectation and clarity of fee entitlement relative to provision of services. These objectives would, however, seem more likely to be met in a relationship founded on specific activities being delivered to an accurate programme, as required by the PSC. Clients are understandably more focused on results than simply on their consultants being active and very few clients are likely to be entirely comfortable with draw-down of fees against time elapsed, as opposed to against complete 'deliverables'. This is a very important distinction for architects responsible for fee bidding and managing projects and they are likely to find more assistance in reassuring clients on this issue through use of the PSC than with the RIBA documents.

Minimum quality of consultants' performance is non-negotiable in the legal context of 'acting with reasonable skill and care', as required by both the PSC and the RIBA standard appointment documents. This clearly relates to design skills, but clients are also just as concerned about the time and cost of performance of professional services as of construction. Competence is critical, but arguably a minimum achievement level, not a goal; clients are rarely 'delighted' by the performance of their architects where there has been any disagreement over the contractual relationship. The PSC would seem to offer an excellent contractual framework within which architects' design, co-ordination and project management skills can be successfully exercised, but only if the same combination of creativity and care is applied to assembling the contract documents as to the project execution itself. Any formulaic tendencies might be better served by the RIBA standard appointment documents.

Table 7.4. *Overview of standard provisions*

Form of contract	RIBA SFA/99	RIBA CE/99	RIBA SW/99	PSC 98
Full consultant design for Employer	✓	✓	✓	Core clause 21
Partial consultant design for Employer	+DB1/99	+DB1/99	–	Core clause 21
Partial consultant design for Contractor	+DB2/99	–	–	Core clause 21
Consultant design includes design by subconsultant	+SC/99	–	–	'Nested' PSCs
'Binding' contract programme	–	–	–	Core clause 31
Performance-related payment incentivization	–	–	–	Main options A and C
Bonus for early completion	–	–	–	Secondary option X6
Delay damages	–	–	–	Secondary option X7
Binding time and cost quotations for change	–	–	–	Compensation Event procedure
Provision for specific partnering arrangements	–	–	–	Secondary option X12

The ability to engage other design team consultants to perform related professional services on the same project under their own back-to-back PSC appointments is clearly a major benefit and arguably much more likely to achieve a 'butt-jointed' totality of professional services for the project, rather than the all too common phenomenon of 'gaps' or 'overlaps' between consultants' services and the resulting delays and expense in their resolution. There is perhaps an opportunity with the PSC to run the project in 'fast-forward' virtual reality, in order to prepare appropriate appointment documents to facilitate execution of the real project.

PSC compared and contrasted with ACE

Bill Weddell

8

The published documents

The Professional Services Contract (PSC) is published for the Institution of Civil Engineers by Thomas Telford Publishing. The various forms of agreement published by the Association of Consulting Engineers (ACE) are under the general heading of 'Conditions of Engagement 1995'. The ACE is a trade association for engineering, technical and management consultancies.

Details of the published documents are given in Table 8.1.

Table 8.1.

PSC	ACE
The Professional Services Contract (second edition June 1998)	Agreement A Consulting Engineer as Lead Consultant
	Agreement B Consulting Engineer not Lead Consultant
	Agreement C Services for Design and Construct Contract Contractor
	Agreement D For Report and Advisory Services
	Agreement E Consulting Engineer as Project Manager
	Agreement F Consulting Engineer as Planning Supervisor
Guidance notes and flow charts	

Notes: (a) Agreements A, B and C are published with two variants: (1) Civil and Structural Engineering and (2) electrical and mechanical services in buildings. (b) Separate documents are published giving guidance on completion of agreements.

Much of the content of the various ACE agreements is common to all ACE documents. The PSC allows various types of appointment by selecting one appropriate main option and suitable secondary options. The Scope in the PSC (equivalent to 'brief' in the ACE agreements) is drafted to cover the parties' precise requirements.

The PSC includes a sample form of agreement in appendix 1 of the guidance notes. The ACE agreement comprises three parts:

- memorandum agreement
- conditions of engagement
- schedule services.

The version of the ACE agreements used in this chapter for the purposes of comparison with the PSC is agreement B(1) 2nd Edition dated 1998.

Use of the documents

The foreword to the PSC states that it is designed for the purpose of appointing professionals to carry out the various roles in the NEC contracts, and wider use of appointing professionals where the New Engineering Contract (NEC) is not used, or even where no construction work is required. Hence the term used in the PSC for one of the parties is 'Consultant' rather than 'Consulting Engineer'. The other party is named as 'Employer' in the PSC and 'Client' in the ACE. The ACE agreement on the other hand is designed for the appointment only of consulting engineers in their various roles in construction.

The PSC with its non-adversarial approach claims that it is particularly suited for the use with partnering agreements. It has been designed for international use. The secondary options allow for the incorporation of clauses required by legislation of specific countries. In the case of the UK, two options have been included – Y(UK)1 and Y(UK)2 – and a further option has since been issued [Y(UK)3] in connection with the Contracts (Rights of Third Parties) Act 1999. The ACE agreements appear to be designed only for work in the UK.

Definitions

The ACE agreement has 23 definitions and the PSC has 11 definitions in the core clauses. Further definitions are given in the main option clauses of the PSC, mainly 'Prices' and the 'Price for Services Provided to Date'.

The PSC specifies the services to be provided in terms of the Scope. The ACE agreement defines the services in terms of professional civil/ structural engineering services and 'work elements'.

The parties' obligations

The standard of care to be exercised by the ACE's consulting engineer in performing the services is that of 'reasonable skill, care and diligence' (B2.3). The PSC requires the Consultant 'to use the skill and care normally used by professionals providing services similar to the *services*' (clause 21.2). It also exempts the Consultant from obeying an employer's instructions that might require the Consultant 'to act contrary to his professional code of conduct'.

As in other contracts in the NEC family, the PSC includes procedures designed to discourage confrontation and minimize disputes. In clause 10.1 it requires the parties to 'act as stated in the contract and in a spirit of mutual trust and co-operation'. Specific matters for co-operation are dealt with in clause 23.

Under ACE agreements, the duties of co-operation are more specific to the consulting engineer's task. He or she is required to 'co-operate in the co-ordination and integration by the Lead Consultant of the design of the Works with that of the Project' and in performing the services to 'co-operate with the Lead Consultant and any Other Consultant' (B2.8).

Subconsulting is permitted in both forms of contract. The procedure in the PSC includes the need to submit a proposal for subconsulting to the Employer for acceptance. The Employer is able to exercise some control over the terms of the subcontract (clause 24.3).

Subconsulting under the ACE agreements is expressed in rather different terms. The consulting engineer may recommend to the client:

- subletting to a specialist subconsultant (B2.5) and
- that detailed design should be carried out by a contractor or sub-contractor (B2.6).

In both cases consent of the client is not to be unreasonably withheld.

Under the PSC, the Employer has a duty to provide 'information and things' as required by the contract and the accepted programme. The ACE agreements are less specific, but much more demanding in that the client is to provide 'all necessary and relevant data and information ... in the possession of the Client, his agents servants, the Lead Consultant, Other Consultants or Contractors' (B3.1). The client is further required to give and procure assistance from the above parties 'as shall reasonably be required by the Consulting Engineer' (B3.2).

The PSC contains a number of obligations intended to promote effective management of the contract. These include specific requirements on communications – for example, their form and period allowed for reply.

The ACE agreement in clause B3.3 specifies that the client should give decisions, consents, etc. 'in such reasonable time so as not to delay or disrupt the performance of the Services'. The PSC also includes an early warning procedure (clause 15) which obliges a party to notify the other party of any matter which might affect cost, result in delay or affect the usefulness to the Employer of the services. This procedure has the effect of highlighting problems early, such that they can be resolved in a manner acceptable to both parties. The nearest equivalent of this procedure in the ACE agreement is in B6.7 which requires the Consultant to advise the client of any additional work which may be required or of any disruption.

Timing and programme

The PSC provides a completion date by which the Consultant must complete the services. The ACE agreement does not provide a completion date for the services. It does however require the consulting engineer to 'use reasonable endeavours to perform the Services in accordance with any programme agreed with the Consulting Engineer'. No express sanction is included in the contract if the consulting engineer fails to complete the services in time. The PSC on the other hand includes an option X7 that entitles the Employer to delay damages for failure to complete by the completion date.

As in other NEC contract forms, the programme in the PSC is an important management tool. Its detailed requirements include the following:

- dates and timing of the work by the Employer and others
- dates for provision of information by the Employer, and
- method statements and resources proposed by the Consultant.

Three of the compensation events listed in clause 60.1 mention the accepted programme. Thus some of the dates on the programme acquire a certain contractual status resulting in possible additional payment to the Consultant if his or her costs are thereby affected.

The ACE agreement refers to a programme only in B2.9. There are no detailed requirements specified and it seems that its main use is for the client's information.

Determination, suspension, disruption and delay

Termination and its consequences are set out in clauses 94 and 95 of the PSC. Under these clauses the Employer has a right to terminate if they no longer require the services. The right of the Employer to terminate at will can only be exercised if secondary option X11 is included in the contract. The ACE agreement includes this latter as a basic right in B5.2.

The ACE agreement in B5.2–5.8 states the right of either party to determine the contract or suspend performance. Similar provisions are included in the PSC, although these are less specific and are stated in lesser detail.

Payment

In the PSC there are four types of contract available in the four main options that determine how the Consultant is paid. The ACE agreement allows for three methods of payment as shown in Table 8.2.

Table 8.2.

PSC	ACE
Option A Priced contract with activity schedule (a lump sum contract)	Lump sum fee
Option C Target contract	
Option E Time-based contract	Time-based fees
Option G Term contract	
	Percentage fee

The PSC caters for two types of payment not provided for in the ACE agreement. The target contract is a time-based contract that provides an incentive to minimize costs. The term contract is a 'call-off' type of contract in which the Consultant provides services only when instructed by the Employer.

The percentage fee basis of payment in the ACE agreement has not been included in the PSC. The reasons for not including this *ad valorem* basis of payment are stated in the PSC guidance notes. They are:

- there is no incentive to produce an economical design/service
- cost of professional services is not related to cost of construction
- final cost of construction is not established until completion of construction
- effect of variations of Scope/brief are difficult to assess.

While the ACE agreement endeavours to deal with some of the above points, the basic objections remain. In particular, the pricing of variations or disruption in B6.7 is by payment for additional work and resources as time-based fees. The 'works cost' is the final cost of the completed works that will incorporate any varied work. Thus the calculation of fees due is not straightforward and may be the subject of some dispute.

The ACE agreement distinguishes between normal services and additional services with different methods of payment for each. These are described in detail in the schedule of services which makes use of the work stages in the Royal Institute of British Architects (RIBA) Plan of Work. Because of much wider scope of services that may be in a PSC contract, no such detail is included. Expenses in both contracts are dealt with in various ways as payment additional to that of the fees.

Intellectual property rights

Under both the PSC (in the core clauses) and the ACE agreement, consultants/consulting engineers retain the intellectual property rights over the material that they produce. But in both contracts, the Employer/client has a right/licence to use the material for limited purposes. However, option X9 (Transfer of rights) in the PSC transfers the Consultant's rights to the Employer who in turn gives certain limited rights over the material to the Consultant. The exercise of this option is a matter initially for the Employer and then agreement/negotiation between the parties.

Liability, insurance and warranties

Both contracts allow for limiting the liability of the consultant/consulting engineer to an amount stated in the contract.

The PSC requires the Consultant to provide three insurances:

- professional indemnity insurance
- public liability or third party insurance
- employer's liability insurance.

The ACE agreement requires only the first two of these, since in the UK, employer's liability insurance is a statutory requirement.

The PSC has an indemnity clause under which the Consultant is required to indemnify the Employer against any infringement by the Consultant of the rights of others. The ACE agreement includes a net contribution clause. It also provides for limiting the period within which actions against the consulting engineer may be commenced. This period of liability is stated in the agreement.

The ACE agreement includes an option regarding the exclusion or inclusion of liability in connection with pollution or contamination (A10).

Both contracts provide for collateral warranties. In the PSC this is in the form of secondary option X8 and in the ACE Agreement it is in B8.7.

Disputes

Both contracts specify that disputes are to be resolved by adjudication. Where under the PSC, the contract is subject to the UK's Housing Grants, Construction and Regeneration Act 1996, it is necessary to include option Y(UK)2 which amends the contract to comply with the Act. The ACE agreement is silent on the resolution of disputes which may arise in agreements which are not 'construction contracts' as defined in the Act. Any reference of a dispute in a 'construction contract' is to be referred to adjudication using the Construction Industry Council (CIC) Model Adjudication Procedure (B9.2).

The PSC allows for a dissatisfied party to refer the dispute to the tribunal which may be described in the Contract Data as either legal proceedings or arbitration. The ACE agreement is silent on any further proceedings, although the CIC Model Adjudication Procedure refers to a possible final determination by legal proceedings or arbitration. Thus the ACE agreement does not appear to include a specific arbitration agreement. It does require the parties to 'attempt in good faith to settle any dispute by mediation'. No specific mediation procedure is included.

Quality

The PSC, in section 4 of the core clauses includes specific requirements on the quality of the services to be provided by the Consultant. They require the Consultant:

- to operate a quality management system as stated in the Scope
- to provide a quality policy statement and
- to provide a quality plan.

This is doubtless designed to give some assurance to the Employer on the quality of the Consultant's services.

Section 4 of the PSC also includes a procedure for notifying defects in the services (defined as non-compliance with the Scope) and for their correction within certain time limits.

Compensation events

The PSC lists 11 'compensation events' which are events at the financial risk of the Employer and which may entitle the Consultant to additional (or in some cases reduced) payment and extension of time. These events are described in detail and some are related to the dates on the accepted programme. The nearest equivalent of these in the ACE agreement is in B6.7 entitled 'Variation or Disruption of Consulting Engineer's Work' where the events are described in more general terms.

Summary

In broad terms it can be said that the PSC is more flexible than the ACE agreement and that it is suitable for a much wider range of consultancy appointments as well as for international work. Its procedures comply with the principles of modern project management. The obligations of the parties are clearly defined with remedies provided for unsatisfactory performance. It gives the Employer much greater involvement in the provision of the services.

The ACE agreement because of its limited use related to construction work, makes drafting of the services to be provided much easier than in the case of the PSC largely because of its prescriptive character and link with the RIBA Plan of Work.

PSC compared and contrasted with RICS

Richard Honey

9

Origins and appropriateness

The New Engineering Contract (NEC) Professional Services Contract (PSC) and the Royal Institution of Chartered Surveyors (RICS) Form of Agreement and Terms of Appointment have distinctly different original philosophies and, as a result, are substantially different in their form and content. The RICS appointment was drafted solely for the appointment of quantity surveyors in the UK. It is part of a guide for clients, and includes the form of enquiry, schedule of services, fee offer, form of agreement and the terms of appointment (the terms).

The PSC may be used to appoint any consultant for any project-related work in the UK or internationally. It is a rather more sophisticated document than the RICS appointment, intended to require pro-active management and to generate a non-adversarial relationship. The PSC, like the NEC, is a demanding contract in terms of the administration required. An appointment under the PSC is therefore likely to be more costly than one under the more straightforward RICS terms, even if this is not reflected directly in the fee charged.

That the NEC structure forms the basis of the PSC is both its strength, when used on NEC family projects, and its weakness when it comes to its wider use. Much of the PSC's structure and text was adopted from the main NEC and, whilst not redundant in professional appointments, some clauses deal with time/cost/quality conflicts which are far less common with consultants than they are with contractors (see the provisions on the programme, quality management systems, defects, early warnings, and compensation events). Likewise, most of the secondary option clauses will be rarely used in UK appointments, save for collateral warranties (X8) and employer's agent (X10), as well as the Y options (compliance with UK legislation). Modifications to the PSC to enable it to be used as a sub-contract are given in appendices 3 and 4 of the PSC guidance notes.

For many projects where in the past the RICS appointment was used, the PSC will not be wholly appropriate and may not only alarm more traditional quantity surveyors, but be positively counter-productive unless both parties are familiar with the contract's rigorous requirements.

Key areas of comparison

Four preliminary points are worthy of note. First, while many of the PSC's provisions are not replicated in the RICS appointment, only the RICS appointment prevents assignment without the consent of the other in writing and enables the Employer to instruct suspension of the services. Secondly, neither of the contracts includes an exclusion of the effect of the

Contracts (Rights of Third Parties) Act 1999, as is now common, although stand-alone Option Y(UK)3 can be incorporated with the PSC. Thirdly, the PSC may be used with the NEC partnering option (X12). Fourthly, there is no form of agreement in the PSC document; it is to be found in the PSC guidance notes. The RICS appointment incorporates a form of agreement. Table 9.1 sets out comparisons of the more minor PSC clauses with the RICS' provisions.

Fee basis

The PSC contains four main option clauses for remuneration which, following the structure of the main NEC, are rather confusingly labelled A, C, E, and G (lump sum with activity schedule, target cost, time charge and term contract, respectively). Option C shares the price risk between the Employer and the Consultant and is a complicated approach to remuneration. While it may help a non-adversarial, co-operative approach to the Consultant's appointment, it will significantly add to the cost of contract administration, given the contract's accounting requirements and the calculation of the Consultant's share. The share ranges and percentages can be used to produce incentivization and innovation, but consultants may not be keen to take the risk of paying money back to the Employer. Option G is a mix of time based remuneration and lump sum prices for specified tasks. Where a task is instructed which is not on the agreed task schedule it is treated as a compensation event; this should ensure that the task schedule is comprehensive. Although one can see the application of such an option within framework agreements, a consultant's tasks will vary so greatly between projects that lump sum prices for tasks would seem to be inappropriate.

The RICS fee offer allows for fees to be by percentage of construction cost, a lump sum, a mix of percentage and lump sum fees for different services, or on a time charge basis. It does not provide for target cost or term-based remuneration, although the RICS terms themselves could apply to fees calculated on any basis. On all but the time charge basis the Consultant is taking the primary risk that the fee will cover the work required.

The PSC does not provide for a percentage fee basis, as the drafters rejected the link between construction cost and the value/price of services, despite this being a traditional method for calculating quantity surveyors' fees.

Payment

Under the PSC the Consultant assesses the amount due and submits an invoice for that amount, with such supporting information as the Scope requires. The amount due is essentially the price for services provided to

Table 9.1. *Comparison of PSC clauses*

PSC	RICS	Commentary
12	Form of Agreement	The statement of the governing law in PSC 12.2 and Contract Data (CD) 1 is mirrored in the RICS Form of Agreement
13	Terms 14	RICS Terms 14 on notice mirrors PSC 13.1, 13.2 and 13.6 in part, but goes further including deeming of delivery (Terms 14.2) and a provision on reckoning of time (Terms 14.3)
20	Terms 2	The RICS appointment is substantially similar to the PSC in its effect
21	Terms 1	Whilst the RICS appointment's comparable provision is notably slimmer in its expression (Terms 1.1) it appears to match the PSC in effect
24	Terms 3.2	The PSC contains provisions on the use of subconsultants, mainly for the Employer's benefit. The RICS appointment precludes subcontracting without the written consent of the Employer
26	Terms 2.1	The PSC covers access to people, places or things; the RICS appointment only covers access to information
70	Terms 6	The PSC governs the use of the parties' materials; each retains ownership but allows use (Option X9 facilitates transfer of ownership). The RICS appointment has an exclusion of liability for use of materials beyond that for which they were prepared. The PSC includes a non-disclosure provision
81	Terms 5	The PSC provides for the Consultant to have professional indemnity (PI), and public and employer's liability insurance. The RICS appointment only expressly requires the Consultant to use reasonable endeavours to maintain PI insurance, and only if available at commercially reasonable rates. RICS Terms 5.2 matches PSC 81.2 in respect of the Consultant
82	Terms 13	PSC 82.1 allows a limitation on liability for a failure to provide the services to be entered in CD1. The RICS appointment includes a limitation to the amount of the PI insurance required, as well as a net contribution clause and a time limitation on actions (see Form of Enquiry 9)

94	Terms 8	Under the PSC either party can terminate on notice for an act of insolvency; the Consultant can terminate for 11 weeks' non-payment; the Employer can terminate for consultant failure or if he no longer requires the services. PSC Option X11 allows the Employer to terminate for any reason, subject to paying 5% compensation. Again, under the RICS appointment, either party can terminate on notice for an act of insolvency. The Employer can terminate on 7 days notice, and the Consultant on 28 days notice if the Employer is in material breach
95	Terms 9	Under the PSC, on termination the consultant ceases work and may be required to assign the benefit of any subcontract. Final payment is made as soon as possible, covering the amount due and any reasonable forward costs; where termination is due to the Consultant's failure or insolvency, the cost of completing the services can be deducted. The RICS appointment obliges the Employer to pay the amount due, any reasonable proportion of the next instalment, and, unless termination is due to consultant default, any necessary consequential costs within 28 days. Both forms ensure certain rights persist
X8	Terms 7 Enquiry 10	PSC X8 obliges the Consultant to enter in to collateral warranties listed in the CD; presumably this does not extend to any others not listed. The RICS appointment uses the British Property Federation warranties; a limitation in number and assignment is also provided for, and the Consultant is not obliged to provided them where insurance cover is not available
X10	Terms 2.2 Enquiry 1	PSC X10 provides that the employer's representative acts with the authority stated in the CD. The RICS Form of Enquiry allows for details of a representative and Terms 2.2 require the Employer to advise in writing of any appointment or change of representative
Y (UK)[1]		The RICS appointment makes no provision in respect of the CDM Regulations; Option Y(UK)1 only relates to CDM and Compensation Events
Z	Enquiry 10	The RICS Form of Enquiry 10.1 provides for amendments to the British Property Federation's forms of warranty agreements, but for no other additional clauses or amendments

[1]Where there is no comparable provision in the RICS appointment nothing is noted.

date plus expenses. The Consultant decides the first assessment date; CD1 sets the subsequent assessment interval. The Employer pays that part of the invoice they agree with, and the Consultant either corrects the invoice to that amount or provides further justification. Payment is made within 3 weeks, or such other period as is stated in CD1. Interest is due on late payments, at the rate stated in CD1. The PSC allows payments in different currencies. It should be noted that the last assessment date under the PSC is 8 weeks after the defects date (stated in CD1); this may be considerably later than a consultant is used to waiting for his or her final payment.

Secondary Option Y(UK)2, where the PSC is used in the UK, deals with the requirements of the Housing Grants, Construction and Regeneration Act 1996 (the Act) by amending PSC 51.1 and adding PSC 56 and 60.4. The only substantial changes are to require a notice from the Employer if they intend to withhold payment against the amount otherwise stated to be due, and PSC 60.4, which provides that the statutory right of suspension for failure to pay under the Act is a compensation event.

The RICS appointment incorporates the payment provisions of the Act, and is therefore substantially similar to the PSC with Option Y(UK)2. The amount due is the total of the instalments of fees and charges stated in fee offer 4 for the services then performed. The key difference is the RICS appointment's adoption of the amount in the Consultant's invoice as the sum due in the absence of a payment notice from the Employer. The RICS appointment provides for interest on late payments and a right to suspend for a failure to pay, as required by the Act.

Obligations and performance

The PSC relies heavily on the Scope document to detail the Consultant's obligations, but leaves its production to the Employer in each case. The Scope should contain fundamental definitions and provisions, including the services to be provided. The RICS appointment uses different parts of the same document in order to define the Consultant's obligations, including the form of enquiry and a tick-list schedule of services.

The PSC contains an enforceable completion date, which requires not only that the work stated in the Scope is done but that any defects which would have prevented the Employer from using the services are corrected. The completion date may be changed as a result of a compensation event. The RICS appointment does not provide for a completion date, but terms 4.6 do envisage that there may be contractual time limits.

The PSC's obligations as to programme, quality management system and defects are not reflected in the RICS appointment. While these provisions may promote active management, they also add to the cost of the services, defeat flexibility in their provision, and may lead to additional compensation for a consultant where it is not otherwise usual. PSC 41 as to defects merely puts in to express terms that which might otherwise be implied.

Compensation

Core clause 6 of the PSC sets out the compensation event procedure. The Consultant is to be compensated for a long list of events, some of which are frequent occurrences in professional appointments and would not normally give rise to additional remuneration. PSC 61.1 requires a compensation event to be notified within 2 weeks of either party becoming aware of it. There is no express provision for what happens if this 2-week deadline is not met, and it is arguable whether it precludes the Consultant from recovering compensation. A failure to give an early warning may reduce the compensation to be paid. There is a procedure for quotations for compensation events. Compensation events are calculated on a time charge basis, together with any assessed delay to the completion date. If the Employer decides that the Consultant has not correctly assessed the compensation event they may make their own assessment, which is then imposed on the Consultant.

Under the RICS appointment the Consultant is only paid in addition to the agreed fee in two circumstances. First, under terms 4.7 and 4.8, where work additional to that contracted for is required due to changes in the scope of the construction works or programme, dispute resolution proceedings with the contractor, or instructed changes to the Consultant's services. Additional fees are calculated on a time charge basis. Secondly, RICS fee offer 1 provides for a lump sum fee to be adjusted by a fair and reasonable amount where the final construction cost varies beyond a specified percentage from that set out in the form of enquiry. In the RICS appointment there is no completion date and therefore no formal adjustment to the time for performance.

Dispute resolution

Although PSC 10.1 obliges the parties to act in a spirit of mutual trust and co-operation, there are no informal dispute resolution procedures in the PSC – only adjudication and arbitration (but see the Y(UK)2 matter of dissatisfaction). The RICS appointment provides for a complaints procedure and discussions in good faith to seek to agree a settlement of any dispute.

There are two schemes of adjudication under the PSC: contractual adjudication under the main clauses PSC 90–92, and statutory adjudication under the Act, implemented by Secondary Option Y(UK)2, which amends PSC 90 and 91. The adjudicator is appointed under the NEC Adjudicator's Contract, and should be named in CD1, along with a back-up appointing body.

Contractual adjudication under PSC 90 is subject to time limits, and is final and binding unless reviewed by the tribunal. PSC 91.2 allows a matter also disputed with a subconsultant to be referred to the same adjudicator at the same time; if the PSC is used as a subcontract, then the Employer may bring the Consultant's dispute into the main contract adjudication.

Option Y(UK)2 introduces a procedure for dealing with a matter of dissatisfaction, including a meeting between the parties, prior to a dispute crystallizing. There is some doubt that this process is compliant with the Act. PSC Y(UK)2 90.11 allows the adjudicator's decision to be overturned by agreement as well as by the tribunal.

The RICS appointment provides for adjudication in compliance with the Act. The Chartered Institute of Arbitrators (CIArb) is the appointing body, and the adjudication is to be conducted in accordance with the Construction Industry Council's Model Adjudication Procedure.

PSC 98 allows for the reference of disputes to a tribunal, which is named in CD1. It may be the courts, arbitration or another procedure. If it is arbitration, the particular procedure to be employed should be stated in CD1. The intention is that adjudication is a condition precedent to the right to refer to the tribunal, and only then within a time limit after adjudication concludes or fails (PSC 98.1). PSC 98.2 appears to provide that the tribunal's decision is absolutely final and binding.

The RICS appointment contains arbitration as the tribunal, and places no limit on the right to refer a dispute to arbitration. The parties are to agree the arbitrator within 28 days, or apply to the CIArb for an appointment. The arbitration is to be conducted in accordance with the Construction Industry Model Arbitration Rules.

NEC/PSC compared and contrasted with PPC2000

Frances Forward

10

Introduction

The objectives of this chapter are first to compare the basis of the New Engineering Contract (NEC) family of documents and the ACA Standard Form of Contract for Project Partnering (PPC2000) and secondly to analyse those aspects which display significant differences in their rationale and operation. This comparison is between the main industry-recognized standard forms of partnering contract and, uniquely, allows the NEC to be measured against an even newer form. In the nature of partnering, this chapter covers contractual relationships for both professional services and construction.

Partnering policy

Partnering has existed as a concept in the construction industry for some considerable time, albeit with a wide variation of understanding as to what it means for real projects. Common principles of co-operative team working between clients, contractors and consultants on individual projects and the possibility of longer term relationships for such teams on future projects have become accepted goals. Publication of the Latham and Egan Reports focused attention on the potential benefits of partnering the entire supply chain; the only hindrance to achieving this was the question of how to enforce co-operative team working on an essentially voluntary basis. A practice emerged of drafting partnering charters, signed by the project team as a 'pledge' to their intention to promote co-operative team working. However, while such documents are often proudly displayed, they rely on individual motivation and are not capable of altering the parties' strict contractual obligations when a conventional, often adversarial, standard form contract is signed in parallel. Indeed, there are examples of such contracts being expressly stated as overriding the parallel partnering charter in the event of a conflict, which inevitably can result in the kind of unproductive dispute which partnering is often promoted as discouraging. The NEC family of contracts has always embraced partnering principles, specifically in its requirement for the parties to act 'in a spirit of mutual trust and co-operation' and it is no coincidence, either that Sir Michael Latham endorsed the use of the NEC across the construction industry, or that a number of partnering projects have been successfully completed under the NEC forms.

A fundamental shift in the concept of partnering occurred with the publication in 2000 of the Construction Industry Council (CIC) Guide to Project Team Partnering, the culmination of the CIC Partnering Task Force research across all sectors of the construction industry. For the first time, partnering principles and contractual relationships were seen to be not only compatible, but also capable of being drawn together under a multi-party 'umbrella'. The CIC Guide set the scene with its 'Model Heads of Terms' for the creation of specific forms of partnering contract.

Industry response to the CIC challenge

PPC2000 was the first response to the CIC challenge, published later in 2000 by the Association of Consultant Architects, as a wholly new standard form of contract. The other standard form drafting bodies had also taken an active interest in the CIC Partnering Task Force findings and the supplementary NEC Partnering Option X12 was published as a consultative version in September 2000, following evaluation of the existing NEC family of contracts, which were found to respond already to many of the CIC recommendations. After detailed review of the consultation findings, the first edition of Option X12 was published in June 2001.

Status and structure

While PPC2000 and the NEC Partnering Option X12 are both predicated on the objectives set out in the CIC guide, there is a fundamental difference between them in terms of their status and structure, giving them complementary appeal both within and across the constituent sectors of the construction industry. PPC2000 is a stand-alone standard form of contract in its own right, which encompasses both construction and professional services and requires subsidiary documentation in order to implement it. The NEC Partnering Option X12, in contrast, sits at the secondary option level of the established 'pick-and-mix' NEC family contract structure, below the core and main option clauses and capable of incorporation into both construction and professional services bi-party NEC contracts between members of the supply chain. Figure 10.1 illustrates the conceptual difference between the two structures.

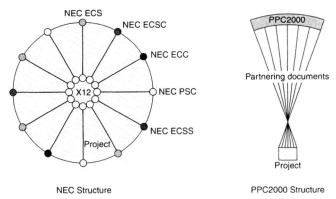

Figure 10.1. Abbreviations:
NEC ECC: NEC Engineering and Construction Contract (black book)
NEC PSC: NEC Professional Services Contract (orange book)
NEC ECS: NEC Engineering and Construction Subcontract (purple book)
NEC ECSC: NEC Engineering and Construction Short Contract (blue book)
NEC ECSS: NEC Engineering and Construction Short Subcontract (turquoise book)

Procedure

Table 10.1 lists the comparative documentation necessary for operating the two forms of partnering contract.

Ownership

Experience with partnering projects suggests that a prime factor for success is the motivation of individual project team members to take joint ownership of the project. The project must come first, but not at the expense of the employers of those individual team members. Partnering should be neither an excuse for exercising commercial power through coercion; nor a reason to forego legitimate contractual expectations. It follows that a key requirement of partnering arrangements is to facilitate the assumption of collective responsibility for the project, ideally on both an interpersonal and a contractual basis. The word 'mutual' in relation to trust and co-operation in the NEC core clause 10.1 and under 'Working together' in the NEC Partnering Option X12 encapsulates this principle and the NEC Partnering Option X12 requires each partner to nominate a representative to act for it, emphasizing the role of individuals. The maintenance of the bi-party NEC relationships ensures that joint ownership of the project engendered by the incorporation of the NEC Partnering Option X12 is complemented by the familiar reciprocal actions, and where necessary leadership, under those bi-party contracts. There is a conscious reliance on the inherent dynamic between the bi-party contracts to ensure that the project moves forward. The multi-party relationship created by the use of PPC2000 should encourage strong feelings of collective responsibility, at both an individual and a corporate level. However, the corollary to the existence of a single contract is the need to allow the potential for external steering of that entity, both in a facilitative capacity and conceivably to combat inertia, if none of the multi-party 'coalition' is actively moving the project forward. The PPC2000 decision to follow the CIC guide's recommendation to appoint a Partnering Adviser and the NEC decision to avoid the creation of such a separate role is a direct consequence of this strategic difference in contractual structure.

Role of the Partnering Adviser

The Partnering Adviser performs a number of crucial functions in PPC2000, which broadly fall into two categories. First, the Partnering Adviser is responsible for ensuring the consistent and timely execution of the supporting documentation which is essential to make PPC2000 operable and secondly, the Partnering Adviser is charged with managing the partnering relationships within the partnering team, including assistance in potential dispute management. Separation of these functions is unnecessary with the NEC Partnering Option X12, as the individual NEC

Table 10.1. *The comparative documentation necessary for operating the two forms of partnering contract*

Form of contract	PPC2000 ( ) = priority of documents	NEC 95 / PSC 98	+X12
Agreement	✓ (i) standard form within contract	✓ (model forms)	
Pre-possession Agreement	✓ (effectively a letter of intent)	No – first bi-party contract suffices	
		Contract Data Part One	
		Contract Data Part Two	
	Price Framework (x)		Option X12 Contract Data
	Partnering Charter (v)	Core Clauses	
	Partnering Terms (ii)	Main Option Clauses	
	Joining Agreement (ix)	Secondary Option Clauses	Including X12
	Key performance indicators (KPIs) (xi)		Schedule of Partners + KPIs
			Schedule of Core Group
	Consultant Services Schedules and Consultant Payment Terms (vi)	Scope	
	Project Brief (vii)/Project Proposals (viii)	Works Information	
		Site Information	
	Commencement Agreement (iv)/Partnering Timetable (iii)		Partnering information
		Programme	

bi-party contracts are managed by the two particular parties and the chosen core group of partners manages any documentation universal to all partners, e.g. keeping the schedule of partners up to date. Relationships and disputes are managed through the operation of the bi-party contracts, incorporating the consistent and universal Option X12 clauses.

Liabilities and remedies

Perhaps one of the reasons that partnering has often been participated in on the basis of a 'voluntary' partnering charter and a parallel 'old-style' contract is that there has been an inherent reluctance to disturb the relatively well-established English law doctrine of privity of contract and the relatively predictable consequential boundaries of contractual and tortious liability. It is not that either PPC2000 or the NEC Partnering Option X12 have ignored such important considerations, or of themselves altered them. Both documents have been drafted in an innovative commercial era and in the context of English law, a potentially altered legal era, with the enactment of the Contracts (Rights of Third Parties) Act 1999. Equally, both PPC2000 and the NEC family of contracts are intended to be operable under other jurisdictions, where different liabilities and remedies exist. The position with NEC is that remedies remain within the dispute resolution provisions of the Partners' own bi-party contracts, with the ultimate sanction of losing any future partnering role. The position with PPC2000 is that the remedies are set out in the dispute resolution section of the partnering terms, but that a number of the liabilities will be considerably variable, depending on the content of the subsidiary documents forming part of the partnering contract. Notably, the Consultant Services and Cconsultant Payment Terms will have contract-specific liabilities attached, as they are an inherently bespoke part of the standard form.

Application

Critical to both the PPC2000 and the NEC Partnering Option X12 methodologies is the principle that potentially the entire supply chain, whether consultants or contractors and whether designing, managing or constructing, can join the partnering team and that their contractual, as well as their interpersonal, relationships with the rest of the partnering team are ordered within such methodology. The two methodologies also share the ability to entertain contributions to the project from persons who are not part of the partnering team, although there are likely to be both philosophical and practical objections to this, unless their contributions are either very small or non-critical. PPC2000 and the NEC Partnering Option X12 are both operable internationally, although this is more readily apparent with NEC, as a result of attempting to place jurisdiction-specific provisions either in chosen secondary options, or in the Contract Data, rather than in the core clauses. While PPC2000 provides for naming the jurisdiction of the contract in the agreement, there will inevitably be some redundant clauses within the partnering terms and the definitions appendix if the contract is used outside common law jurisdictions. There is, however, a precedent for other possibly superfluous clauses in PPC2000, such as the apparent ambivalence between 'complementary' and 'prioritized' partnering documents. Future familiarity may allow PPC2000 to be shortened.

Compatibility

Both PPC2000 and the NEC Partnering Option X12 are inherently incompatible for integration with other standard forms of contract within the partnering team, due to their respective methodologies. PPC2000 simply does not need to be integrated with other forms, due to its holistic approach to the partnering team. However, the recent publication of a 'sister' subcontract, SPC2000, does not entirely square with the original concept of a single multi-party partnering contract. One possible area for caution with PPC2000 is that consultants may be tempted to try and use other standard appointment documents to provide the necessary level of detail for the Consultant Services and Consultant Payment Terms. While this might partly work in relation to simple issues, e.g. work stage definitions, it is likely to create conflicts within the contract overall and it would be a negative use of the Partnering Adviser's expertise to have to deal with the consequences. The NEC Partnering Option X12 augments the existing NEC/PSC clauses and it would simply not make sense if it were 'attached' to any other forms. Conversely, the NEC Partnering Option X12 will not work without an NEC/PSC contract structure behind it, as it is deliberately not designed as a contract in its own right. It may prove to serve a 'stand-alone' role solely as an *aide-mémoire* in focusing early attention on the appropriate nature of a planned partnering project and subsequent contract documentation. Table 10.2 shows some of the parity between the two contractual methodologies.

Table 10.2.

Form of contract	PPC2000	NEC 95/PSC 98	+X12
Standard Provisions			
Partnering Adviser	✓	–	
Client representative	✓	✓	
Core Group of Partners	✓	✓	
Priority of documents	✓	–	
Formal problem-solving hierarchy	✓	✓	
Core Group dispute review	✓	✓	
Early Warning procedure	✓ (Partnering Terms Clause 3.7)	✓ (NEC Core Clause 16 / PSC Core Clause 15)	✓ (Secondary Option Clause X12.3(3))

Conclusion

The construction industry is showing an unprecedented interest in partnering, as the possibilities for tangible commercial benefits have become apparent. While not every client sponsoring a project, or series of projects, will wish to engage in and promote partnering, for the growing number of clients who do wish to partner, the availability of two standard methodologies can only be welcomed. The choice between the NEC Partnering Option X12 and PPC2000 is likely to be made on a stylistic and conceptual level, rather than following any particular sector tendency within the construction industry, particularly as single- or multi-project partnering can be accommodated with either methodology. Clients will possibly be influenced by an almost instinctive preference as between a multi-party relationship and connected bi-party relationships, as well as a desire to equate new appearances with readily recognizable and familiar models. Clients will also be influenced by a number of pragmatic factors, including whether the project team has prior experience of the contract methodology, which is currently more likely with NEC and the PSC, due to their earlier drafting. The ease of partners joining and leaving is arguably greater under the NEC Partnering Option X12 than under PPC2000. One factor which may yet prove decisive as more clients are making the choice, is the need to employ a Partnering Adviser with PPC2000 whilst such a person would be superfluous with NEC and PSC. This clearly has financial resource implications, but it may also have human resource implications, as a result of the sophistication of the documentation to be managed in this role. It is somewhat difficult to imagine a responsible and conscientious client entering into a PPC2000 contract without specific legal advice, whereas the NEC Partnering Option X12 maintains the simplicity of the rest of its family members and is therefore capable of implementation by any competent construction professional.

Lightning Source UK Ltd.
Milton Keynes UK
29 January 2010

149266UK00001B/51/A